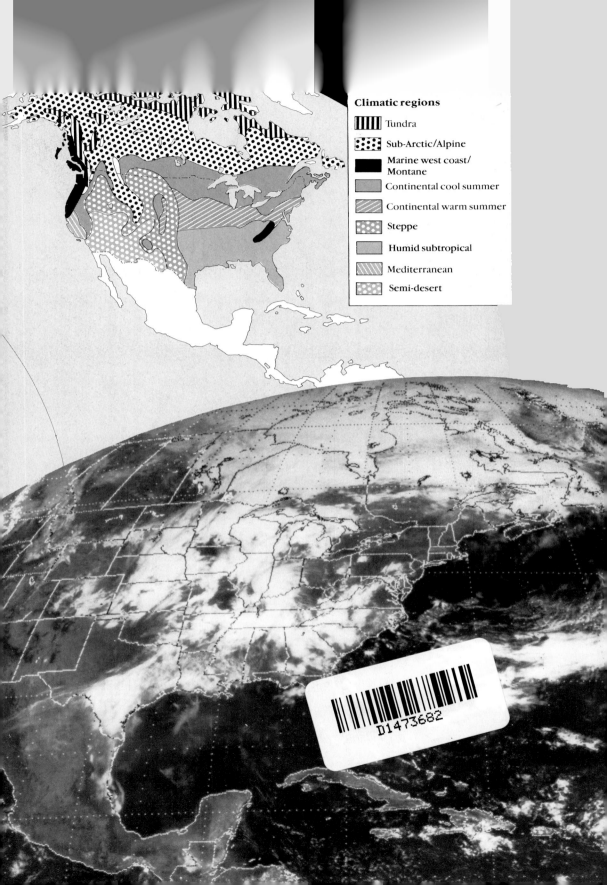

Climatic regions

Tundra	
Sub-Arctic/Alpine	
Marine west coast/Montane	
Continental cool summer	
Continental warm summer	
Steppe	
Humid subtropical	
Mediterranean	
Semi-desert	

☆BELLAMY'S NEW WORLD☆

☆BELLAMY'S☆
NEW WORLD

A Botanical History of America

David Bellamy

BRITISH BROADCASTING CORPORATION

This book accompanies the BBC Television Series
Bellamy's New World, first broadcast on BBC1 during Autumn 1983

Series presented by David Bellamy
and produced by Mike Weatherley

Published to accompany a series of programmes
prepared in consultation with the
BBC Continuing Education Advisory Council

Dedicated to Hans Jenny

This book is set in 11 on 13 point Monophoto Garamond
Printed in England by
BAS Printers Limited, Over Wallop, Hampshire
and bound by Dorstel Press Limited, Harlow, Essex
Colour sections printed by
Cook, Hammond and Kell Limited, Mitcham, Surrey

First published 1983
Published by the British Broadcasting Corporation
35 Marylebone High Street, London W1M 4AA

ISBN 0 563 16561 8

Contents

CHAPTER ONE
Cactus Cops

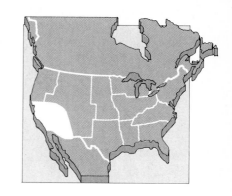

I was standing beside my pickup somewhere along State Route 85 out of Gila Bend, Arizona, minding my own business, enjoying a breathtaking panorama of desert and contemplating the intricacies of crassulacean acid metabolism, when it happened: the wail of an electronic siren, flashing red and orange lights (not quite Starsky and Hutch but the full routine), two well-armed officers of the law. 'May I see your driver's licence, and can we take a look in the pickup?' I agreed to both requests and reached for my Hertz rental pack.

'From the UK,' said one, and turning to his partner, waving my licence, he said, 'Ever see one of these things?'

The answer. 'Nope, but he's clean. Just camping gear and bricks.'

I inwardly hoped that they wouldn't ask about the bricks. Fortunately they did not, but went on to explain that they were on the lookout for cactus rustlers—and that my pickup made me match the description of one of the most wanted sort of men in that part of the West.

You can probably guess how the rest of the conversation went, for it was great to see the real armed arm of the law enforcing legislation aimed at protecting the local flora. They were giving backup support to seven undercover cactus cops led by Richard Countryman, whose job it is to patrol 90,000 square miles of Arizona's desert lands in an attempt to stop the lucrative racket which supplies big cacti to landscape gardeners in the neighbouring states and across the world.

I listened enthralled to tales of real-life car chases and shoot-outs between rival gangs—and came to realise that cactus rustling is a real prickly business in more ways than one.

A pickup not unlike mine roared past at well over the official 55 mph speed limit and my two new buddies set off in hot pursuit, but not before giving me instructions of how to get to the cactus pound.

For a botanist on holiday what better place to start than a desert? And there are none better than those of the south-west of America (for location see light area in map above). The ideal place in which to start a tour of American deserts is at Organ Pipe Cactus National Monument, for there the Organ Pipe and Senita Cacti and the weird Elephant Tree have their only stations within the country.

These three succulents mentioned above typify what is known as the Gulf Coast phase of the Sonoran Desert, which crosses the Mexican border at this point and just extends into the USA. However, two more widespread forms of desert are also well represented within the area of the Monument. In the west the Californian microphyll (which means 'small leaf') desert is dominated by Creosote Bush, Sage Brush, and a whole host of half shrubs. The damper, more upland, east rejoices in one of the most varied arrays of succulent plants to be seen anywhere on earth: Saguaro, Teddybear Cholla, Chainfruit, Prickly Pear and Agave, to name but a few. These together typify the Upland Arizona Desert which has a small rainfall during two periods of the year.

Half shrubs are plants which put on new growth and a good crop of leaves after rain, live it up while they can and then die back to the old woody growth, or even further, during drought. Succulents are plants which usually forgo the luxury of leaves and rely on a thick fleshy stem, which both stores water and, being green, carries out all the functions of photosynthesis. For example, the aptly named Barrel Cactus is a mere shadow of its rotund self during periods of drought, only swelling up to its true dimensions after rain has passed its way, which is very rarely.

Members of many different plant families have adopted half shrubbery and/or succulence, thus maintaining a place for themselves in the deserts of the world. Most have gone further and developed crassulacean acid metabolism as well.

The raw materials of photosynthesis are water and carbon dioxide. To obtain the former, land plants have roots and to obtain the latter, their shoots are well endowed with stomata (pores) which open to let the gas in. Unfortunately, as carbon dioxide enters the plant, water vapour leaves it and that is bad news for any plant which is trying to live in a dry habitat.

To combat this water loss the desert plants only open their stomata at night, when it is cool and evaporation is at a minimum. Their problem then is that, as the carbon dioxide cannot be turned into sugar in the dark, it has to be stored in the form of organic acids. The cell sap of the succulent thus becomes more and more acidic as its night-storage acid batteries become fully charged. Come the sunrise, the plant shuts up its stomata and gets on with the job of turning the stored organic acids into sugar, thereby returning its internal acidity to normal. This self-inflicted dyspepsia goes under the name of crassulacean acid metabolism, a small price to pay for the freedom of the world's deserts.

Please remember that every night when you have turned off the light your pet cactus goes through its own routine of internal acidification for, of all the families of the flowering plants, the *Cactaceae* has got it all off to perfection, despite the fact that the process gets its name from another juicy family, the *Crassulaceae* or Stonecrops. It is of great interest to note that the *Cactaceae* is an all-American family, for its members only grow naturally in the New World. Any true cacti which are found growing elsewhere have been introduced. The native plants of the Old World which look and act like cacti are members of other families, and they provide us with the best example of convergent evolution: that is, examples of unrelated plants which look alike because they have become adapted to overcome certain problems in the same way. It is, however, easy to tell a true cactus from one of its imitators, for its spines are always borne on small raised pads called areolae. It is even easier to tell when they are in flower—and when they are, the desert becomes alive with colour.

'Living' and 'desert' are two words which should not relate, for a desert is, by definition, a place which is too dry, too cold or too salty to support the growth of plants and, hence, is devoid of life. Semi-deserts or arid lands would be more correct, despite the fact that deserts and cacti have become almost synonymous. For the latter reason and to save words I will stick with deserts, but you will know more exactly what I mean.

To accentuate the paradox still further, I can safely say that, if you go for a walk through any of those types of American deserts you will see more in the way of wildlife than on a walk of similar length through the Amazonian rain forest. The reason is not that there is no wildlife in Amazonia but that there is much more which gets in the way when you are trying to see it—you can't see the wildlife for the trees; but you can in the desert, and it is very abundant.

Perhaps the most endearing inhabitant of these arid lands is the diminutive Kit Fox, now sadly an endangered species over much of its former range. Much easier to see is the Jack-rabbit, whose gigantic ears serve as an efficient device both for gathering sound and for dissipating heat, thereby protecting their owner from its two main enemies: predators and the sun.

Both Kit Fox (*left*) and Jack-rabbit are well adapted to desert life.

To avoid the heat of the day, many of the desert animals have gone on the night shift, so a walk in the cooler parts of the day, early morning and late evening, will reveal a diversity of mammals, birds and reptiles on their way either to or from bed. The list is very long and includes the smallest owl in the world, the Burrowing Owl which nests and rests underground, and some very ugly customers, in terms both of appearance and venom, although I reckon that the Diamondback, which warns of its presence with a rattle on its tail, is one of the most exquisitely marked snakes and can in no way be called ugly.

For the less intrepid, a walk even in the heat of the day may be rewarded with a sighting of a Desert Pronghorn Antelope or, most magnificent of all, a Desert

Bighorn Sheep. The latter is just one of ten races of North American sheep which today enjoy the largest and most diverse range of habitat of any living species in its group. Each race, living in isolation on its own range of mountains or area of land, has gone its own woolly way to success, but not so far as to deserve the distinction of specific rank. They look different, behave differently and live in different places, but if brought together they can interbreed. It is, therefore, both correct and best to refer to them as different races.

The Desert Bighorns share with the camels, some varieties of which roamed North America in prehistoric times, a fantastic ability both to tolerate dehydration and, like the cacti, to rehydrate rapidly once water becomes available. During the cooler, wetter season from December to May they derive sufficient water from their diet, which at that time is at its most juicy, and water which is excess to their needs is voided as dilute urine. During the rest of the year, these large sheep— their very size aids their heat regulation by protecting them against rapid gains and losses—drastically alter their behaviour. They rest during the heat of the day, protected in part by their light-coloured thick wool coat. An extra large, water filled stomach complex, not only aids digestion of their intractable diet but also acts as a reservoir, compensating for some degree of water loss. Likewise, their kidneys conserve water by withdrawing it from the urine, the salts in which can reach concentrations more than four times that of sea water.

Summer showers may produce a renewal of juiciness in their feed which may help to stave off the long-term effects of drought, but finally, when body weight has fallen by up to a staggering 20% a visit to a water-hole becomes a matter of survival. The fact that each visit is made as brief as possible, that rehydration is very rapid, and that no animal has been observed to over-indulge, indicates, that throughout their evolution, visits to water-holes must have been fraught with extreme danger. In recent times their major predator has been man, but in the not too far distant past both Lion and Sabre-toothed Cats must have lain in wait at the water-holes.

Apart from the cacti and other succulents and half shrubs, all of which must be regarded as drought tolerant, there are other ways in which plants on which the bighorn is dependent have maintained themselves in the desert environment. Many, like the Palo Verde and the Smoke Bush, both of which are extra tall by the standards of desert plants (3 metres or more), only inhabit the beds of arroyos which flood with water after heavy rain and maintain a wet subsoil for much of the rest of the year. They are characterised by deep roots which follow the water down, tapping all levels of the soil during the dry cycle. The native Fan Palm, *Washingtonia*, cannot grow in such places; its shallow roots restrict it to oasis areas in which springs keep the surface layers moist throughout the year. A knowledge of the local botany can, therefore, lead you—and the bighorn— to water and some plants can even provide you with a drink. In dire need it is possible to chew the succulent stems to obtain water, but great care must be taken to avoid a mouthful of spines, and of salt, for many concentrate the latter.

The most abundant plants, at least at certain times of the year, and without doubt the most stunningly beautiful, are the ephemerals. These avoid the hot dry summer by the not-so-simple expediency of enduring it in the form of dormant seeds. The mechanisms by which the seeds remain dormant, and then emerge from their dormancy, are many and varied. For this reason, each shower of rain brings with it the excitement of seeing what exactly will pop up! There are always the old faithfuls, which come up every year and they really should be called 'annuals'. But there are the others, and there's always the hope of an entirely new one. Yes, even experts who have made a lifetime's study of one area, still encounter new species with surprising irregularity.

The conditions which govern the emergence of seeds from dormancy include single-threshold factors and combinations of conditions and circumstances, like correct temperature, hydration, light of varying wavelengths; these must in some cases be preceded by scarification or chipping of the seed by sand grains moved by wind or water, or even damage by insects or passage through the digestive tract of a whole variety of animals. However, once the specific trigger or triggers have been squeezed, the floor of the desert becomes a new living experience.

The living desert of Arizona should not be missed, and I must admit that I was loath to leave its arid pastures to make my way across to California. But first, I just had to visit the cactus pound. There I met Richard Countryman himself, who showed me the collection—and the ones recently brought in certainly looked a miserable bunch. He indicated the signs of maltreatment: deep gashes on the base of an Ocotillo where a chain strung from the bumper of a pickup had been used to drag it quickly from the ground; a black exudation where damage had caused necrosis of a big Barrel Cactus. All this is good proof of 'cactus bashing', for a licensed dealer would take his time and avoid damaging such a valuable cargo. They were now in good hands, under intensive care for, even when uprooted with every precaution, the spreading net of surface roots will be damaged, and only correct handling, once replanted, can ensure survival. The cactus crook is aided by the fact that it may take a large specimen more than a year to die and actually look dead; by which time he is well away from the irate owner, who may have paid between $3,000 and $4,000 for his seven-metre crested Saguaro status symbol.

The long-term inmates of the pound were all doing very well and there were quite a lot of them, because it may take several months to prove or disprove the fact that they have been taken illegally. The problem is that there are official dealers licenced to take choice specimens for resale from private land or land which is under development—and any law of licence is open to abuse. If a felony is proven, the cactus crook is shaken with a fine or is even stirred (the local term for going to jail), and the cacti are confiscated, to be used in municipal landscape projects of benefit to the local society. In most cases they cannot be returned to the wild because, in the absence of special aftercare, they would not survive a second transplant, however carefully it is done.

Take the mighty Saguaro for example. It is the biggest of the bunch so you might expect it to be the toughest; but no such thing. For a start, once moved the Saguaro must be planted in exactly the right place, a rocky slope with adequate but not too much drainage being the most appropriate. Then it must be planted exactly the right way round with regard to the sun; if not it may get sunburned. In this hot, harsh climate a few degrees out of line can make all the difference for, as it grows it produces a thicker cuticle on the south-west, that is the sunnier, side. Extreme care must also be taken when manoeuvring a big one into position; arms break off very easily, and broken spines leave wounds through which disease can enter. In fact, you should handle a big cactus like an American breakfast egg, easy over and sunnyside in the right direction. Then it must be treated with rooting hormones and given the right amount of water over the long period of reestablishment: too much and it will grow long and spindly and may topple over; too little and it will dry out and die.

Growth rate, even in the most suitable desert locations, is exceedingly slow: 75 years to a first branch and a hundred to make six metres. Then it can live to a ripe old age, 250 years, and grow to a height of twenty metres, by which time it will weigh over six tons and may have produced as many as twenty million seeds. The Saguaro blooms throughout May and is pollinated by Long-nosed Bats and White-winged Doves. Unlike most bats, which eat insects, these sip nectar; and the White-winged Doves carry a golden hood of pollen from flower to flower. Many other birds get in on the act as do honey bees, but these were only introduced onto the pollinating scene during the last century.

From seed to 'sapling' is the most critical phase in the life of any plant, and the same is true for the cacti. The desert harbours, or rather dry-docks, a ravening horde of animals who delight in Saguaro seeds and, especially, seedlings before they have put on their full armoury of spines. The latter not only keep the browsers at bay but, being shiny, reflect away some of the sunlight and so protect the plant to a certain extent. They also collect water from fine mist and rain and drip it down to the hub of the extensive root system, which infiltrates the surface soil for tens of metres in every direction. This method of water collection is especially important when the plant is at its most thirsty. Then it is also at its most flaccid, and all the spines will be pointing down to the base, concentrating the drip feed system just where it can be most rapidly effective. The fluted Saguaro stem acts like a concertina or bellows so that the stem can rapidly expand as it refills with water.

Meanwhile, back to the seeds. The ones with the best chance of survival are those which fall in, or are carried to, a shady place; for germination in the full sun will be lethal. Shade may be provided by rocks or 'nurse' trees such as Mesquite or Yellow Paloverde, beneath which the seeds are often dropped by birds with bad table manners. The same shade can also protect them from winter frosts—and the searching eyes of herbivores, who are always on the look-out for a ready supply of food and water, both of which are provided by succulent seedlings. So it is that only very few, on a long-term average one per adult, make it and raise their spiny presence to maturity. Throughout their long lives they play an important part in the life of the desert, each one a focal point. They are chewed at and burrowed into by a host of insects and birds. They provide safe nest-holes for wood-peckers responding to the wound by shutting it off from the precious water-storage tissue with a thick woody scab layer. When they die, their rotting tissues provide food for another group of plants and animals, the arboreal decomposers. Eventually all that is left are the internal woody rods which provided the plant with the strength and the meagre supplies of water it needed for its magnificent life, and the woody casts of each woodpecker hole. The former look like a frozen fountain as they spread out and the latter can provide visitors with a most unusual souvenir, a portable hole all of their own!

I learned much about cacti and conservation from the cops with green fingers for, above all the people I have ever met, they understand that their region will only remain rich if they can stamp out cactus rustling, and protect this their natural heritage. After profuse thanks I went on my way again, west across the high country that forms a border between the Colorado and Mojave deserts. Like the rest of America, it is not a place to hurry through; for the bizarre rock formations not only change at every turn of the road, but with the passage of the sun across the sky. As everywhere, sunrise and sunset can be breathtaking, but even the near-vertical mid-day sun casts its own shadow set across each heat-shimmered scene.

I just had to stop and see the Joshua Trees in all their glory—and there was a road sign pointing to another National Monument of that name. They were in flower and, before I entered the monument area, I saw a specimen which had been knocked down by a car so I stopped to sample a bloom. I took just one from a massive creamy-white spike of many hundreds. It was just as the books describe, a perfect example of a flower of the Lily family, its floral parts, petals, anthers and carpels arranged in whorls of three. Carefully checking that it hadn't been damaged by a Yucca Moth, I popped it into my mouth, for the book also said that they were eaten by the local Indians. 'Joshua, Joshua, sweeter than orange squash you are': I don't think that that was how the song came to be written because it was a somewhat bitter mouthful, but I do know that it was the Mormons who gave this weird tree its name, for when it is mature it looks just like a man holding up his arms to God.

All around me in this area was another plant in full yellow flower, called Mormon Tea, a member of the class Gneteae, a group of plants which is, in fact, a non-missing missing link. They bear minute flowers on minute fleshy cones, and thus bear some resemblance and perhaps some relationship to both the flower- and the cone-bearing plants. Their stems and tops can be roasted and then brewed into tea and also were used as a cough mixture. The fact that its Latin name is *Ephedra* and the anti-congestant Ephedrine was obtained from a close relative which grows in Asia emphasises the fact that some folk medicine is founded on scientific fact.

Joshua Trees are Yuccas, to be exact *Yucca brevifolia* and Yucca Moths visit their flowers in search of nectar and to lay their eggs. In so doing they pollinate the flowers and thus provide the larvae of their progeny with food, for they eat

Abundant creamy flowers of the Joshua Tree attract Yucca Moths to their nectar.

the developing seeds. That is why I made a careful inspection before eating mine.

My route then took me north-west through the Great Basin, which is not so much a basin, more a way of life. It is a gigantic upland area, criss-crossed with valleys and mountain ridges, and is itself flanked by the Sierra Nevada on the west and the Rockies in the east. The high mountains produce a rain shadow area in between, and the general topography and aridity of the area determines that, although melt waters do flow down from the snow-clad peaks each spring and after rain, the total inflow is insufficient to produce an outflow. So the various depressions within the Great Basin act as sumps, into which water containing dissolved mineral salts drain, but from which water is lost only by evaporation. The lakes found within the area, and there are many, are therefore all saline, as the name of the biggest, the Great Salt Lake, proclaims.

In the not too distant past, when the world was in the grip of the last ice age and the high peaks of the local mountains were covered in glaciers and ice-fields, spring melt kept the basins much better supplied and the lakes were both much larger, and their waters sweeter. In fact, much of what is present-day Utah and western Nevada was submerged beneath a vast inland sea. I was on my way to one such shrunken lake, which is situated close by the boundary of the Yosemite

National Park and is fed by spring melt waters from the peaks of the Sierra Nevada itself. Its name is Mono Lake and its future is much tied up with my story of the cactus cops.

I had my first glimpse of Mono Lake from route 395 out of Reno, just south of Conway Summit. To the east was Cedar Hill, all that remains of a million-year-old volcano, the peak of which was once an island sticking up in the glacial lake. Then 40% of the whole catchment basin was covered in water, in places to a depth of 300 metres. From my vantage point, I descended across a series of old beach terraces, cut as the ancient lake waters dried and receded. To prove it, here and there amongst the sagebrush weird grey rocks could be seen, which looked as if they had bubbled up out of the earth. Closer inspection between the encrusting lichens revealed a covering of crystals of thinolite. These rocks are tufa and were, in fact, bubbled up out of the ground while the area was still part of the floor of Mono Lake.

At its greatest ice age extent, the lake level stood at around 2188 metres above sea level and covered an area of 89,355 hectares. When visited by Mark Twain in the last century it had reached a new equilibrium between inflow and evaporation, and its shoreline stood at 1953.8 metres and its saline waters (5.2% salinity) covered a mere 2,274 hectares. Standing at that level today, you have to walk a muddy kilometre and a half and drop another eleven metres or more to find the contemporary water level and, when you do, you find it has a salinity of 9.5%. Apart from the lovely sticky mud, the walk takes you through a great variety of tufa formations. These range from gigantic towers and spires, which appear to be constructed out of concrete cauliflowers, to intricate fretted bluffs and curtains of rock worked into the weirdest of patterns.

The former were formed where sub-aquatic springs bubbled up through the lake bed into its saline waters; the latter, where similar springs seeped through saline sands before discharging into the lake. As the lake level has dropped, the sand has been washed and blown away, to reveal their presence and then to continue their erosion. The exact method of formation of these tufas is both varied and complex and centres on the peculiar chemistry of the Mono Lake waters, which are charged with salts, soda and sulphates. The latter two constituents have been put there by volcanic activity and all three concentrated by evaporation. There is, in fact, so much soda in the water that all you have to do is wade fully dressed into the lake, which has a distinctly soapy feel, to get your trousers fresh laundered on your legs.

Tufa formation is part chemical, brought about by the mixing of different solutions and, where conditions allow the growth of plants, it is part biological. All living things produce carbon dioxide as a by-product of their life processes; if it is released into water which is already charged with dissolved salts of calcium, calcite will be formed and will come out of solution. Much of Mono Lake's tufa is made of calcite and within its towering structures can be found impressions of the microscopic plants which helped in its formation, each frail impression encased in rock.

The lake itself is thus surrounded by an almost lunar landscape with the added spice of hot springs and volcanic cones and craters. It is a place of great wonderment, only approached elsewhere by parts of the Afar Triangle in the north of Africa's great Rift Valley. The most wonderful thing is, however, the lake itself for, despite its tripartite bitter saline and sulphurous makeup, its waters teem with life. The raw materials of plant growth—water, carbon dioxide and dissolved mineral salts—are there in abundance; the latter, in fact, in too great abundance and very few plants and animals are able to tolerate life in its concentrated waters. For those that can it is a paradise.

In winter the lake waters are fully mixed by wind and stand at a cool $0°$ to $5°$ Celsius. The microscopic plant life reproduces rapidly at this time and with nothing to eat it, becomes abundant enough to colour the water a deep green. So it remains throughout the winter, very little sunlight being able to penetrate into the depths through the algal bloom.

Spring sunshine at last begins to warm the lake, upper layers first and, as cold water is denser than warm water, the lake becomes stratified into a lighted upper part warmed to a temperature of $10°$ to $22°C$ and a cooler darker bottom zone which remains at around $5°C$. By the end of March the upper water is alive with the larvae of Brine Shrimps which have hatched from eggs lying on the bottom of the lake. The larvae feed on the microscopic plants. Eight to twelve weeks and fourteen moults later they have become adult shrimps, each one like an animated feather just over one centimetre long. They then spawn and set a second generation on its way, which by June has all but grazed the plant life out of the water, which in consequence is no longer green. Underwater visibility now approaches ten metres or more.

As autumn gets underway the Brine Shrimps begin to die off but not before they have laid their eggs, which sink to the bottom of the lake, each one protected by a tough cyst. In the absence of their grazers, the plants start to multiply once more, colouring the water green ready to start the whole cycle off again.

Mono Lake is not quite a monoculture for, apart from the Brine Shrimps, its shallow waters also produce, at certain times of the year, the best behaved flies in the world. To a sunbather the background buzz of the Brine Flies which swarm along the shore is enough to send him or her rushing for shelter or the spray can. There is no need, for the Brine Flies don't bite, they don't even bother to land on the bathers' statistics, however vital! All they do is buzz along the beach and lap up the microscopic plant life on the strand line as they go. They are, in fact, the most fascinating things to watch. When the right time comes, the gravid female fly walks along on the surface of the water until, finding a convenient piece of protruding tufa, she envelopes herself in a bubble of air and, with great difficulty, clambers beneath the surface carrying the air and her eggs with her. After laying the eggs, a few dozen at a time, she simply lets go and floats back to the surface, to pop out of her bubble as dry as a fly should be.

The eggs hatch out into wriggling larvae which crawl about over the tufa, grazing on the plant life as they go. After three moults, which take in all some six to seven weeks, the larvae clamp down on the rocks and turn into pupae within thick-walled puparia. Three weeks later, when they have turned into winged adults, the puparia fill with air and float to the top, where the adult flies emerge to get on with their job of spoofing the sun-worshippers.

Now for the statistics: at peak density, one cubic metre of lake water can contain 50,000 Brine Shrimps; dry weight of the total population is estimated at around 3,000 tonnes. As for Brine Flies, 40,000 per square metre of shore over many kilometres is a conservative estimate. With all this potential food on the leg and wing, it is little wonder that just about all the North American species of shore bird, 79 in all, come to feast on Mono's shores, and with no fish to compete for the food, they can gorge themselves all day.

Almost one-fifth of the world's population of California Gulls nest on the islands on Mono Lake, in April crossing the Sierra Nevada from their wintering quarters on the coast. 100,000 Wilson's Phalarope arrive on their way south in late July to spin upon the water, the lobed toes of their feet drawing a vortex of Brine Shrimps up to be scooped off the surface. The females come first; then the males who have incubated the eggs; and finally the young: all on their way to over-winter in Argentina and Bolivia. By the middle of August they are gone, their place taken by 20,000 Northern Phalaropes, again on their way south for the winter. They, too, go their way, leaving the lake to a million Eared Grebes who also stop over on their long flight south.

All this and so much more is there, to be seen and enjoyed year by year. If Mono Lake dried up, all this would be gone and the ecology of a large slice of the world, which includes the full range of each of these birds, would change. Why, even the great whales of the southern oceans benefit from the bounty of Mono Lake, for Northern Phalaropes have been seen picking parasites from off their backs! But surely a lake which has outlived the changes brought about by at least three ice ages, and the warm dry periods in between, will not dry up!

Ever since man has been on the Mono scene he has had more than a passing interest in its waters. Within historic times the local Kuzikin Indians came each autumn to harvest the pupae of the Brine Flies. They were dried in the sun, de-shelled, and the oily protein-rich mini-goujons were eaten, or carried away to be traded with other, less nomadic, local tribes. Adventurers, trappers, loggers and miners came and went their nomadic way, and sheep ranchers settled the sweet-water meadows along the inflow streams, and grazed them into dusty extinction. Each left their own distinctive mark on, or account of, this extraordinary place.

Then in 1907 an Irish immigrant, one William Mulholland, saw the potential of the Los Angeles region, if only sufficient water could be found and tapped. Six years later, Mulholland's ditch, 386 kilometres long, brought the first water from Owens Valley and, when it first gushed through on 13 November, 1913, the lucky Irishman said, 'There it is, take it.' Los Angeles did just that and, by the 1930s, 1.2 million people lived within its sprawling metropolis and money was available to extend the water net to tap the inflows of Mono Lake.

It took 1,800 men six years to tunnel their way, braving noxious volcanic gases

Metropolitan Los Angeles: its insatiable demand for water threatens Mono Lake.

and streams of boiling water, under the Mono craters, and the first water flowed
in 1941, sapping the life blood of the lake.

As far as water engineers are concerned, it is a perfect system; downhill all the
way, no pumps, no expense, in fact it generates its own hydro-electricity on its
way. Like Topsy, Los Angeles just grew and grew, and Mono Lake shrank and
shrank to its present level and crowded way of life. Fortunately its rich wildlife
lived on.

But now there are plans to divert still more of its water supply. Already, islands
on which the birds have, to date, nested and rested in peace are now mere peninsulas, their rookeries open to predation by coyotes and the like. The lake level has
already dropped by 13.7 vertical metres, and it is planned to drop it another 15,

when it will be only one fifth of its natural size. Its salinity will then stand at around 27% and its soda concentration will be enough to rot, not wash, your socks.

Then there will be no need for the gulls to try to nest, nor for the migrants to stop over, because the shrimps and flies will be dead. A vast living resource will have become no more than a chemical sump. Winter will no longer see the rejuvenation of the lake by overturn, but the same winds will whip up enormous clouds of alkali dust, posing problems for wildlife and people alike across a wide area. It can't happen!

But it already has. Owens Valley, which was, in living memory, good grazing country surrounding a similar lake, is today a dust bowl, a hell-hole of little use to man or nature.

But the fight is on. Concerned citizens, many from the metropolis itself, are working together, not only to put a stop to further water diversions, but to reverse the process. They know that, to save Mono Lake, Los Angeles must cut its water consumption by 85,000 acre feet a year—a staggering amount. Impossible, say some voices of authority. Well, in the drought year of 1977, with very little hardship, by watering their lawns only in the evenings, sweeping instead of hosing dead leaves from the sidewalk, swimming in full, not overflowing, pools, they conserved 97,000 acre feet of the then precious water, more than enough to save Mono Lake.

The Mono Lake committee proposes similar, much less stringent measures, which would, over a number of years, save the situation. It will cost money, say some ratepayers, but already their money is being spent in alkali dust control, and more will be required as the grey-white clouds spread their health hazard even further.

It is a complex business and there are many sides to any argument. Yet, I believe that the battle will be won, for this is affluent America.

To return to the bricks, with which this chapter started. I had coated each one in plastic so that they could cause no damage, and everywhere I went in my trip across the Great Basin, I left a present: two bricks in each of the sumptuous five to six American gallon flush tanks on the toilets. Yes, the Americans' proud boast of the largest of everything is upheld even in the rest room! Only a small gesture, you may say, but if everyone in California did the same, the future of Mono Lake and its vibrant community of life could be saved.

The reason that I have much more than hope is that, on many occasions, as I lifted up the lid of the cistern, I found that I had been beaten to it: two bricks, or some other device known in the local trade as F.R.T.Ds (flush reducing toilet dams) were already in place. What is more, even in Beverley Hills the lush English-type gardens with extensive lawns, all of which need lots of irrigation, are being replaced by—what else?—cacti. Hence the recent upsurge in the illegal cactus trade. Well, you can't win them all!

But we can, and we must, win consistently, for the future of mankind depends on it, as I hope this book will show.

CHAPTER TWO
The Real Estates of America

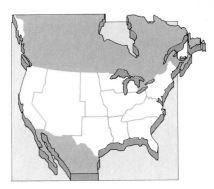

*E*VERY day more and more people across the world start their day with a slice of America. America is the world's biggest producer of wheat, from which derive breakfast cereals, bread, pasta, cake, and biscuits, and of corn products, which are used in the making of tortillas, the manufacture of many animal feedstuffs and most processed food.

Although it isn't true to say that America feeds the world, it is a fact that if these two American mega-crops failed, the spectre of mass starvation would soon become a reality. At present the stability of the whole human world is dependent upon two crops produced by one country: a remarkable fact, especially when put into the perspective of history. It was only in 1776, a little over 200 years ago, that the American Declaration of Independence freed the colonists from oppressive ties with the Old World and set the American Dream in rapid motion: a dream of humanity and the freedom to develop the natural resources of a continent; a dream which came to fruition and put one Neil Armstrong on the moon from which point of human vantage he could look back at Space Ship Earth, the life systems of which then supported some 3.6 billion hungry human mouths. Through the swirling clouds of water vapour which give this ill-named planet Earth its uniqueness within the solar system, its potential to support life, he could see the outlines of America, the country whose broad acres were supporting both his mission and much of the hope of that world. The map at the head of this and the other chapters shows the USA divided into the main farm production regions.

How did all this come about? What is so special about that great slab of real estate which supports both the affluence and the influence of the United States of America in what are today the most crucial of world affairs?

The answers lie in an understanding of its natural history: how it got its size, shape, landscapes, plants and animals; and above and below all, how its soils developed, and how man came to inherit and develop their potential.

The only problem is, where to start such a complex and important narrative. For want of a better, I have chosen as trigger-point an earth-shattering event, a date, a place and a new beginning.

On Thursday 20 March, 1980, a seismometer in a basement in Seattle gave a decisive twitch, a signal picked up by other instruments across the world. It was the start of a chain reaction of events which 59 days later culminated in an explosion of immense magnitude, its epicentre not far from the township of Cougari, Washington State, USA, an explosion which devastated an enormous fan-shaped area, stripping and flattening full-grown Douglas Fir trees 32 kilometres away; covering whole valleys in moving mud, and inundating a vast area with a choking blanket of 150 million tons of toxic dust, great columns of which rose skywards obscuring the sun and threatening the well-being of farmers trans-state, trans-continent and trans-world.

Newspapers, and all the electronic might of the media, carried the messages and images of destruction to every nation on Earth and all the sophistication of twentieth-century technology sat back in the awesome knowledge that nothing could be done to stop the holocaust.

After lying dormant for 123 years, Mount St. Helens had blown her top, or at least 6.7 billion tons of it, proving once again that the seeming timeless tranquility of earth's broad landscapes is only skin deep. Beneath the crust of solid rock which supports us and the four million different sorts of living thing with which we share this planet, there lurks a molten magma, pent up, waiting to demonstrate its power: a power which can both make and break mountains, create oceans and move continents; and much more.

Engulfing rivers of mud and debris stream from erupting Mount St. Helens.

No sooner had the first dust of destruction settled on the flanks of the volcano and across the world than the process of natural reconstruction had begun, a process which has helped to ensure the continuity of life on land for almost 600 million years—its product: living soil.

The products of all volcanic action are igneous, that is, fire-born rocks which may be recognised from all others by their make-up, sparkling crystals which were formed as the molten magma cooled and solidified. If their formation took place deep underground where heat loss and hence cooling would be slow, the crystals are large and readily visible to the naked eye, as in the dense, heavy granites. Lighter basalts composed of much smaller crystals form closer to the surface, where cooling was more rapid. The crystals in the light clinker-like lavas and dusts are so small that they cannot be seen without the help of a microscope: they were formed at the surface where air cooling was very rapid.

All the rocks which go to make up the earth's surface started their life in this way and all are made up of the same rock-forming minerals: olive green Olivine; black prismatic Augite; black prismatic yet strangely fibrous Hornblende; the commonest of all, colourless greasy Quartz; whitish Feldspars; flaky glassy Micas; Garnet; Tourmaline; and many others. The rarest, such as diamonds, are much prized today as precious stones and a ton of igneous rock must be processed to yield half a carat of Diamond, even from good deposits.

The others and especially the first six in my list may not be as precious but are of much greater importance, for being so common and abundant they make up the Lithosphere, that is, the surface of the earth.

The rock-forming minerals are in turn made up of various admixtures and combinations of the 98 stable elements. The same is true of you and me; our bodies are no more nor less than chemical elements borrowed temporarily from our environment to which they will be returned: volcanic earth, volcanic ashes, volcanic dust. There they may begin the whole cycle of life anew, one part of the great cosmic system which works to one set of rules:

Matter and energy, which are but two forms of one and the same thing, can neither be created nor destroyed;
All systems, be they alive or dead tend to run down, that is to disperse the energy associated with them.

The commonest of the 98 elements which form the substance of the rock-forming minerals and hence the face of the earth are, in order of abundance, Oxygen, Silicon, Aluminium and Hydrogen.

Oxygen we all know something about for we need it in the air we breathe. Hydrogen is the lightest element on earth and exists bound to oxygen in rocks and in that unique combination, water, which is not only the commonest substance, covering 70% of this ill-named planet, to an average of 4,000 metres, but also is the universal catalyst which speeds all chemical reactions on their way and without which life would be impossible. Silicon and Aluminium are two elements which make light of modern life: the former as glass (and in microchips); the latter compounded into alloys.

Of every one hundred atoms that go to make up the rock-forming minerals, sixty are Oxygen, twenty Silicon, six Aluminium and almost three are Hydrogen. That is how common they are. They are compounded upon cooling into complex crystals, aggregations of which went to make up the choking dust cloud spewed out by Mount St. Helens.

When rain first fell on the new ash fields it set in motion two processes of erosion —mechanical and chemical—which will continue as long as the rock minerals remain exposed to the influence of water, wind and life.

The first and most spectacular depends on the fact that wherever rain falls upon this earth, it possesses a certain amount of potential energy which, obeying the universal rules, will be dissipated as it flows under the influence of gravity down to the sea. At first it is a gentle process, the insistent blow of falling raindrops; then illution, a downward press as it soaks into cracks and amongst particles of every size. From that point on it gathers momentum, concentrating its effect into seepages, springs, streams and eventually into the power of great rivers.

If you want to see and feel this force of erosion for yourself, there is no better place to go nor thing to do than ride the boats from Lees to Pierce's Ferry down the Colorado River as it flows through the Grand Canyon, the grandest feature of a land which boasts the biggest and best of everything.

It is both a mind- and body-bending experience one of the greatest six-day rock spectaculars the world has to offer. Each rapid, and there are no less than 288 to make or mar your days, provides a new experience of the destructive power of water, while each languid stretch, each eddy-filled pool and alcove, and each vista of layered rock, speaks of the creative power of that same element when combined with time.

For me the greatest experience of all must be the rapid succession of falls, Hall, Granite, Boucher and Crystal, which lead to one of the few places on earth in which time itself has stood still: the inner sanctum of the gorge where granite almost as old as life itself has been exposed by erosion and then polished to perfection by the press of passing water and the scour of the products of erosion it contains.

Reds, yellows and blues, each shimmering in its crystalline perfection, greet the eye at every turn, bluffs of Schist intruded by dikes of Granite which under immense pressure had flowed liquid into their position: igneous rock, the firm basis of it all; firm until attacked by water, now a whirlygig of new experience.

The rubber raft lurches, shoots up and begins to jack-knife on the crest of a roller-coaster wave. 'Hang on', 'Tighten your lifejacket', and other rapid shouts of 'Surprise' as a ton of water slaps you in the face. And behind it all, the noise of rocks grinding upon rocks. A muted cacophony of power which cleaves boulders out of cliffs and then reduces them in time to elemental form—pebble, sand, silt—until they disappear finally into solution. It is not all lost for at each stage in this trail of destruction something is created, for each reduction in size is paralleled by an increase in surface area and hence in the intimacy of the intermix of rock and water and hence the potentialities of solution. Likewise, at each stage in the process, as the mass of the particles is decreased each will be lifted and carried by a lesser pace of current.

My mind is wandering. The boat lurches and jack-knifes me out into the water.

All around is foam and turbulence. Which way is up? My lifejacket drags me to the surface amongst walls of white water. The experience is over almost as soon as it began. I float buoyant; three strong strokes and I am in the slack waters of a pool and all around me the sediments sink in order of their mass to the bottom of the pool—new sediment, which will be scoured and lifted on its way when the Colorado flows in its full winter fury.

Looking up from the depths of the canyon, the layered rocks tell the same story again and again, each layer recounting its own tale of the power of water which made the sediment and then laid it in position: boulder beds laid down by pulsing torrents of the past; shingle banks formed in deltas or thrown up by wave action on the margin of some primeval sea; sandstones, siltsones and mudstones laid down in the stiller waters of ox bows, lakes, tidal estuaries and within the sea itself, and then squashed into rock form by the weight of new sediments descending from above.

Then there are the purest of sediments, those which were laid down—or is it built up?—as landlocked seas evaporated in the dry heat of climates past. Their order and their chemistry belies the solubility of their constituents, the least soluble, like calcium and magnesium, being laid down first, the most soluble, like potassium and sodium, last. Thus it is that through this process of erosion, deposition, solution and evaporation, elements which form only a minute fraction of the original igneous rocks may be extracted and concentrated by nature.

New volcanic activity pushing magma into and through sedimentary rock may speed this process of change and the concentration of constituents. The rock which comes into contact with the magma will be baked, changed, metamorphosed into a different form. Metals diffused throughout its structures may melt and run together in the form of rich veins and lodes, and a whole new array of compounds and textures may come into being.

What nature can concentrate, it can also disperse, for fine sediments, whether changed or in their natural state, once high and dry can be eroded by the wind to be laid down as Loess in another location. One area's grains thus became another area's Loess, a gain of mineral matter, and so the cycle of redistribution and revitalisation of landscapes goes on as it has since the time that the rocks of Marble Gorge cooled into crystalline form.

Of all the rocks which form the ramparts of the Grand Canyon, the limestones of Bass, Muav, Temple Butt, Redwall and Kaibab, names that will ring in every river rider's memory, tell the story of this vitalisation in greatest detail. Limestones are made predominantly of chalk, which is itself made up of calcium, carbon and oxygen. In many cases these elements have been concentrated from solution in water and compounded into living form (the shells of molluscs and much smaller animals, skeletons of coral, armour of sea urchins, otoliths of fish, to name but a few) before at death they joined the rain of sediments down to the bed of the sea in which they lived.

It is all there, written upon the walls of the Grand Canyon; the evidence of more than three billion years of geologic time—eruption, erosion, redeposition—

laid bare by the might of the Colorado for all to see. Those walls comprise the visible history of the structure of the earth; but much more than that—for a closer look reveals the tell-tale print of life—they display a fossil history, reaching back through all life's own antiquity; from the first real life-forms up to a time when 600 million years ago those seas in which the sediments were laid down teemed with a wide variety of life, the stock in trade of creative evolution, which would in the fullness of time take over all the earth. The traces include fossil bacteria, algae, fungi, protists, jellyfishes and their kin, and a multitude of water-living plants and animals which, lacking woody tissues and backbones, left but a frail imprint in the shales and silts. They are all there, within the Grandness of the Canyon itself, there in sequence, in an ordered timescale of events, now, thanks to the sophistication of science, understood in detail.

Fossil seed ferns near the Kaibab trail, Grand Canyon.

I defy anyone to stand, feet awash in the silty waters of the Colorado river, dwarfed by the spectacle of, and the implicit knowledge contained by, that great cathedral of rocks and doubt the existence of a creative mind. Those towering rocks and the fossils they enshrine are the *vade mecum*, the Genesis to Revelation, the proof, the Bible of creative evolution. They are not the red-herring graffiti of some limited idea of a God who could only see as far as the organism he was creating at the time. They are the ordered message of events by which creative evolution worked his purpose out through an immensity of time.

To think otherwise is bigotry in its purest form, for that same creative presence set man aside from all other products of creation through the power of conscious thought. Any man or woman, be he or she Christain, Muslim, Hindu, Buddhist, or whatever creed who buries that one supreme talent worships not a creative presence but Baal and must cry in vain, locked out by his own inner darkness of incomprehension.

Please, whatever faith you claim to have or to have not, come to the Grand Canyon and walk the great staircase of time up from all our fiery beginnings, through the cathedrals of the rocks, towards Zion National Park. The Kaibab limestone which forms the plateau rimming the canyon itself was laid down a mere 225 million years ago, almost at the end of the journey through time, but just at the start of life upon the dry earth.

The cliffs of Zion span the next one hundred and fifty million years and relay in ordered sequence the movement of life from the seas up on to the harshness of the land where it had to obtain its requirements for mineral salts in the interface between the falling of the rain and the flowing of the river. Amphibians and ferns living betwixt water and land benefited from both habitats; the age of reptiles and of cone-bearing plants came and went; the role of the organisms was replaced at least in part by birds, mammals and flower-bearing plants. The detail and the message of evolution grow stronger and stronger as wood and skeleton, both internal and external, lifted life up into open space and thin drying air. The story then continues up through the younger rocks exposed in a scatter of canyons and sites, all like Zion and Grand now fortunately protected as National Parks and Monuments, up to the youngest rocks of all exposed on Brian Head Mountain, 3,448 metres about sea level. These were laid down when the living systems of the world were much the same as they are today: forests and grasslands in which primates and hoofed animals lived and evolved, reaping the full benefits of all the environments the dry land has to offer.

Standing on Brian Head in a sub-arctic alpine environment which precludes the growth of trees and has its counterpart at sea level in Alaska, many hundreds of miles to the north, it is possible to look back not only through these rock corridors of time, but down across many of the climates which America offers today. The timberline casts its ragged presence across the Markagunt Highlands, a mixture of open swales and rich green meadows with stands of Alpine Fir, Engelmann's Spruce, and Limber Pine. Lower down, around the 3,000 metre mark, Bristlecone Pine, Arborvitae, Hemlock; within the confines of the Zion Canyon and lower down to the Kaibab plateau, Douglas Fir, Ponderosa Pine and Aspen. Each zone parallels a change in climate especially the decrease of rainfall and the rise in temperature. Below the lip of the Grand Canyon, a transition zone—Gambel Oak, Piñon Pine, Mountain Mahogany and One Seed Juniper. Then into the dryness below 2,000 metres: the canyon flats can no longer support the growth of trees, and the hardier bunch grasses, Blue and Wheat, take over, along with some desert shrubs. And in the depths below 1,000 metres the effect of little rain and the oven-like heat when the sun is at its zenith makes for a true semi-desert landscape, supporting only scattered Sagebrush, Ocotillo and Mesquite, the like of which typifies the border lands of Mexico, much further to the south. In all, almost the whole north-east/south-west transect of the American environment is telescoped by altitude into the experience of no more than a three-day hike.

Brian Head: there can be no better place both to start and end my story. The vistas are all around and you can stand there in the knowledge that during the laying down of the last few metres of rock both man and the giant slab of igneous, sedimentary and metamorphic rock which is the continent of North America came to have their separate existences, existences which would in the last few centimetres of this great staircase, cut and shaped by wind, water, and time, be linked by, and come to fruition in, a great nation, whose affluence made it the first to set one of its kind upon another planet, an affluence rooted in the richness of rich soils, the product of eons of erosion.

In the summer of 1927, a Pullman train chartered for the exclusive use of the members of the First International Congress of Soil Science travelled clear across the USA, from Washington, D.C., to the west coast and back. Wherever the train came to an official halt, owner-driven cars were waiting to take the delegates to see the sites, the soils of this great continent. On board, a young Swiss-born chemist sat amongst his mentors and marvelled at the experience. His name, Hans Jenny; his destiny, to occupy the prestigious chair of Soil Science at Berkeley, California. Of that grand tour Jenny later wrote, 'The tour was a thrill and opened a new world.' The red soils of the south and the black soils of Canada were showcases of the theory which states that climate and soil type are inextricably linked.

It must be remembered that back in 1927, which is less than an average human lifespan away, many of the soils of America were still to be opened up to agriculture to provide the burgeoning population both of the New World and the Old with food. It is also cogent to add that the dust bowls and grapes of wrath of the '30s were just around the corner and that Jenny, through his life's work in proving the validity of the climatic theory of soil formation, was to become the champion of wise use of this, America's greatest resource.

It was on this trip that Jenny realised that America was a chequerboard of contrasting climates and rock types which, if viewed in the right way, and studied in a grand enough manner, could provide real answers concerning the genesis, the potential and the husbandary of soils.

He set himself the task and he did just that. By 1930 he had demonstrated a strict mathematical relationship between the amount of organic matter measured as nitrogen present in a soil, and rainfall. At constant temperature, soil nitrogen rises with increasing rainfall. By 1933, he had recognised a similar, yet more complex, relationship between climate and clay. At constant temperature, the clay content of soils increases with increasing rainfall. Likewise, at constant average rainfall, the clay content of soils increases with increasing mean temperature.

You may, in hindsight, say, 'So what?' The formation of both organic matter and clay are chemical processes and as such they must take place in water and double their rate of reaction with every $10°C$ rise in temperature. What is more, the Russian-based climatic theory of soil formation had already been introduced to America and was much discussed long into each night of that epic train journey. So what advances was Jenny to make?

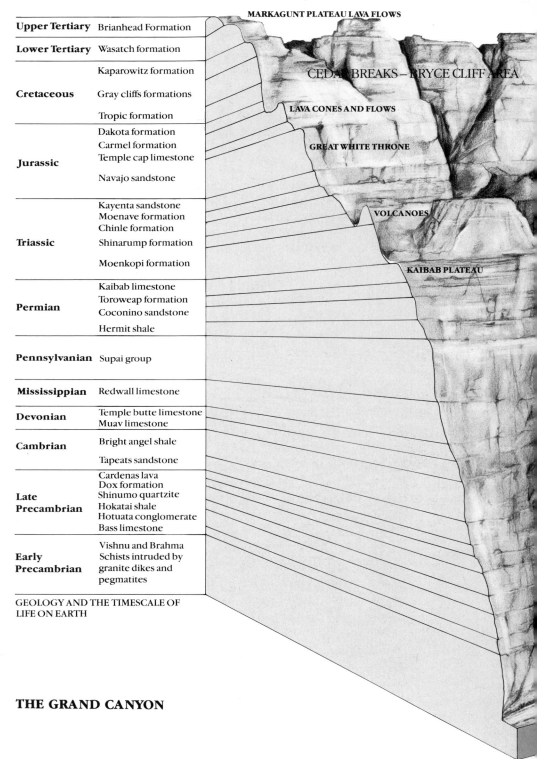

Upper Tertiary	Brianhead Formation
Lower Tertiary	Wasatch formation
Cretaceous	Kaparowitz formation
	Gray cliffs formations
	Tropic formation
Jurassic	Dakota formation
	Carmel formation
	Temple cap limestone
	Navajo sandstone
Triassic	Kayenta sandstone
	Moenave formation
	Chinle formation
	Shinarump formation
	Moenkopi formation
Permian	Kaibab limestone
	Toroweap formation
	Coconino sandstone
	Hermit shale
Pennsylvanian	Supai group
Mississippian	Redwall limestone
Devonian	Temple butte limestone
	Muav limestone
Cambrian	Bright angel shale
	Tapeats sandstone
Late Precambrian	Cardenas lava
	Dox formation
	Shinumo quartzite
	Hokatai shale
	Hotuata conglomerate
	Bass limestone
Early Precambrian	Vishnu and Brahma Schists intruded by granite dikes and pegmatites

GEOLOGY AND THE TIMESCALE OF
LIFE ON EARTH

MARKAGUNT PLATEAU LAVA FLOWS

CEDAR BREAKS — BRYCE CLIFF AREA

LAVA CONES AND FLOWS

GREAT WHITE THRONE

VOLCANOES

KAIBAB PLATEAU

THE GRAND CANYON

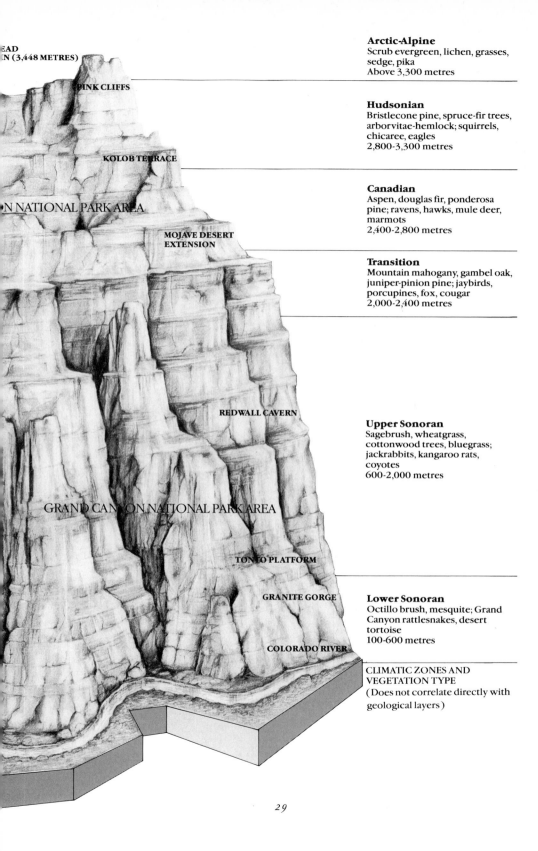

**EAD
N (3,448 METRES)**

PINK CLIFFS

KOLOB TERRACE

N NATIONAL PARK AREA

MOJAVE DESERT
EXTENSION

REDWALL CAVERN

GRAND CANYON NATIONAL PARK AREA

TONTO PLATFORM

GRANITE GORGE

COLORADO RIVER

Arctic-Alpine
Scrub evergreen, lichen, grasses,
sedge, pika
Above 3,300 metres

Hudsonian
Bristlecone pine, spruce-fir trees,
arborvitae-hemlock; squirrels,
chicaree, eagles
2,800-3,300 metres

Canadian
Aspen, douglas fir, ponderosa
pine; ravens, hawks, mule deer,
marmots
2,400-2,800 metres

Transition
Mountain mahogany, gambel oak,
juniper-pinion pine; jaybirds,
porcupines, fox, cougar
2,000-2,400 metres

Upper Sonoran
Sagebrush, wheatgrass,
cottonwood trees, bluegrass;
jackrabbits, kangaroo rats,
coyotes
600-2,000 metres

Lower Sonoran
Octillo brush, mesquite; Grand
Canyon rattlesnakes, desert
tortoise
100-600 metres

CLIMATIC ZONES AND
VEGETATION TYPE
(Does not correlate directly with
geological layers)

What Jenny did was to divine the central truth of the climatic theory, to collect firm data and construct a mathematical relationship which could be put to the test. This is the function of science and there is no more important branch of science than that which concerns soil. He then went on to do much more, for using his training as a chemist he looked into the very make-up of soil particles, giving us a new insight into the intricacies of soil and its formation. All this is published in his principal work, *The Soil Resource*, at the beginning of which he lists and summarises the most important features of the ten main categories of soil, a classification developed by his collaborators, and especially by one of his own graduate students Guy Smith.

In alphabetical order they are:
Alfisols, from the fact that both aluminium (Al) and iron (Fe) are important in their structure. (1)
Aridosols, from the Latin *aridus*, for they are formed in situations where extended periods of drought occur. (2)
Entisols, young soils with little or no structure, soils in the making. (3)
Histosols, from *histos* meaning tissue, for they are rich in the partially decayed parts of plants and form in places which are permanently wet. (4)
Inceptisols, from *inceptus*, beginning, relatively young soils, typical of young landscapes.(5)
Mollisols, from *mollis*, meaning soft, for they are well textured, rich in organic matter and soft to the touch, formed in areas with an adequate supply of water to support natural grasslands.(6)
Oxisols, being highly oxidised, well-weathered soils of ancient landscapes, formed under tropical conditions. (7)
Podsols, from the Russian *pod*, which means ash, for the upper layers are bleached white by the percolation of rainwater, and the lower layers enriched with aluminium and iron. They are formed in pervasively damp climates. (8)
Ultisols, from *ultimus*, meaning last, well-structured mature soils, often with the bulk of the useful minerals in cycle in the woody climax vegetation they support. They are characteristic of warm, well-watered climates. (9)
Vertisols, from *verto*, to turn, dark soils rich in clay which crack on drying to produce wide fissures. (10)
Numbers in brackets refer to the diagram, opposite.
Why name the soil types so precisely? Because until you can name them accurately you can't discuss them with others, and discussion is the most important part of scholarship.
Jenny would be the first to admit that he couldn't have done it by himself without that contact with experts which is the hallmark of any university environment. He would also admit that without the diversity that is America and the affluence of that country which funded his travels, his research and that university environment, he would not have been able to forge the linking synthesis between the macro vision and the sub-microscopic happenings which make soil out of rock.

SOIL TYPES

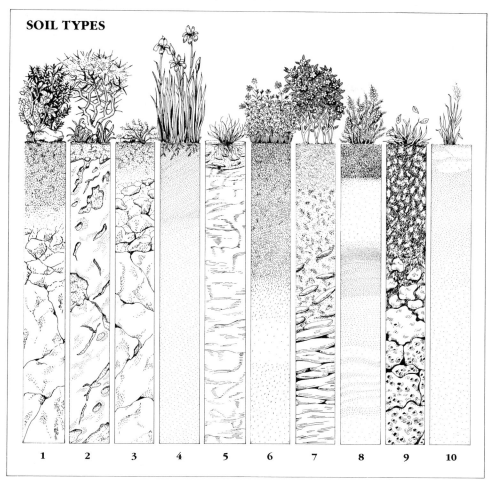

| 1 | 2 | 3 | 4 | 5 | 6 | 7 | 8 | 9 | 10 |

Jenny knew that the active ingredients of soil formation are oxygen, silicon, aluminium and hydrogen, the commonest elements on earth compounded into complex crystals coupled with the commonest compound on earth, water. Under the environmental conditions which pertain over much of the earth's surface, water is a liquid and both silicon and aluminium are insoluble. However, when time and a variety of chemical by-products are added to the intermix of mechanical erosion, solution begins to take place. Water insinuates itself into the molecular structure of the crystals, helped where the temperatures are right, as on the summit of Brian Head, by the freeze and thaw of ice bringing about both a reduction in particle size and a distinct change in chemistry.

The result are new types of minerals called clay, with a particle size below 2μ in diameter, which means that a million could sit with ease upon a pin. The clays thus have an enormous surface area, a factor which is enhanced by their internal make up: each consists of a latticework of layers or plates of silicon and aluminium atoms bonded together by combination with oxygen and hydrogen.

If the structure of these clay minerals were perfect, they would, in essence, be inert, bearing little or no electric charge upon their surface. They are not perfect, for compounded within their structure are lesser amounts of other elements: magnesium, calcium, potassium, iron and many more. Wherever one of these imperfections occurs, the lattice layer structure is left with a minute electric charge, positive or negative depending on the imperfection. These charges, however small, must be neutralised, and they are, by the binding of other elements and compounds within and onto the structure. Thus aggregations of other particles are possible and new structures came into being on the surface of the earth, opening it up to the percolation of water and the penetration of air. Within these structures and these particles a reservoir of minerals is held safe against the leaching power of pure water, but, and here's the magic of it all, available to instantaneous exchange with any charged system of chemistry which comes within its aegis.

We do not know how the chemistry of life came into being, but we do know that one end product of that chemistry which makes plants grow, worms turn, fish swim, birds fly, and man surmise upon the origins of life are free hydrogen ions each with a single positive electric charge. These are voided into the environments, in the case of plant roots into the soil, where they may be exchanged for minerals held within the clay, mineral nutrients which are themselves required to play a part within the living system.

The end result would be an aggregate of clay, organic matter and minerals charged with hydrogen ions and hence sour and acid in reaction, but for the fact that this same acidity can speed the process of chemical erosion and keep the whole complex, which now must be given its proper name, living soil, saturated with other minerals, and hence, fertile.

To unravel this process in full, Jenny needed a field laboratory which matched up to some very exacting requirements. What he required was an area in which the same bedrock and hence parent material for the formation of soil had been exposed over a span of time long enough to see the process through to completion, and under a similar climate. He found it, or at least a close approximation to it, at Jug Handle Creek on the coast of Mendocino County, California, and not a million miles from his laboratories at Berkeley. This whole section of coast is rising up out of the sea and it has been doing that for probably at least a million years. What is more, its rocky foundations have not been tilted, although they have been eroded into cliffs at their seaward face. The result is a series of five broad steps, each one older than the last and each made up of the same sandstone which goes under the name of Greywacke.

In make up, Greywacke averages 17.5 feldspar particles to every 100 quartz grains. The latter, being almost pure silicon dioxide, constitutes but little in the way of minerals useful to plant growth. Within the former, one in every four silicon atoms are substituted by aluminium. The resultant deficiency of positive electrical charge binds sodium, magnesium and calcium (all of which are essential for plant growth) into its structure.

This solid rock is the parent material of which each step is made, material which, once raised above the reach of waves, was opened to the power of both mechanical and chemical erosion mediated by 97 cms. of rain at an annual average temperature of 11.6°C.

The top of the first terrace (the tread of the first step) is no longer made of solid rock, but is covered with a layered soil which is as much as 150 cms. deep. Over one quarter of the feldspar has been weathered away and both clay minerals and organic matter are abundant. The store of available minerals both in the soil and in cycle is enough to support a lush growth of grassland rich in species (more than 100 in all) as do similar prairie soils across the world. The rapid turnover of the annual growth of herbs and grasses feed the decomposers, a rich soil fauna of worms and other creeping crawling things, keeping them active in their job of tillage. Likewise, the abundant organic matter left by the decomposers themselves is slowly being broken down feeding in its turn a whole host of bacteria including those which keep the system charged with the important constituent of living, nitrogen, by fixing it from the atmosphere.

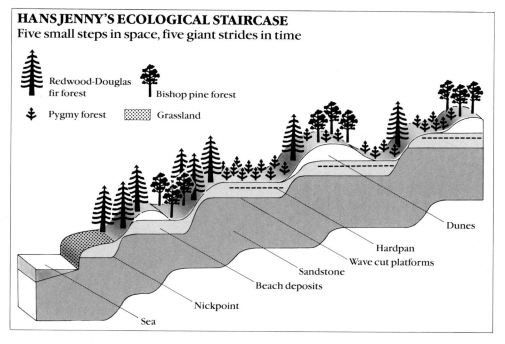

HANS JENNY'S ECOLOGICAL STAIRCASE
Five small steps in space, five giant strides in time

Redwood-Douglas fir forest

Bishop pine forest

Pygmy forest

Grassland

Dunes

Hardpan

Wave cut platforms

Sandstone

Beach deposits

Nickpoint

Sea

There can be little doubt that the salt-laden onshore winds both add to the system minerals and more Greywacke in the form of sand, and at the same time stunt the growth of pioneer trees. The latter must account at least in part for the fact that this first terrace is under grass and not under forest, as it has been since before the first photographs were taken or loggers moved into the hinterland.

The second step, about two kilometres inshore, stands out not only because it is 23 metres higher than the first, but because it is covered with luxuriant mixed forest of Coast Redwood, Douglas Fir, Grand Fir, Western Hemlock and its summit with a forest of tall Bishop Pine. The first four in the list are such good timber trees that they are now planted across the world, the stock in trade of modern forestry. Here in their native home they grow in all their magnificence.

Beneath these mighty trees there is a wealth of other plants (a total diversity of some 70 species) and below them a deep soil profile, rich in clay minerals and organic matter. Bark and twigs and needle leaves, all products of the stately trees and shrubs, are much less tractable to decay and are therefore tackled by a different work force of decomposers, releasing exchangeable nutrients—calcium for building-cell membranes and cell walls, potassium to help in transporting nutrients into and within the living cells, and magnesium to form a part of chlorophyll (all integral parts of life)—which are plentiful, held in the upper horizons of the soil and in long-term cycle in the standing crop of timber and other plants.

This is not, however, the whole story, for on the next three steps the tread, where it is exposed, no longer supports lush diverse woodland, but pygmy forest groves of trees which, though of the same age and in some cases the same species as their giant counterparts upon the lower step, are gnarled and dwarfed almost beyond recognition. Though decades and even centuries old, few rise to a height of more than 3 metres and the majority are much smaller. Bishop Pines and slender Mendocino Cypresses may reach 4 metres, while Bolander Pine, first discovered here in 1822 and only known to grow here, a close relative of Beach Pine, is rarely much more than 2 metres high. Likewise, the few undershrubs (a total flora of only 30 species) which grow hereabouts, Manzanita, Labrador Tea, Salal, Rhododendron and Huckleberry, are all dwarfed and show signs of poor nutrition, disease, dieback symptoms, and infestations of Dwarf Mistletoe, a parasite.

Investigation of the soil shows that something drastic has happened. A cap of raw humus covers an ash-white layer of varying depth. Most of the useful nutrients have been leached out from the upper horizon and down through the soil profile. With them has gone much of the clay which, now mixed with iron, has been cemented together in the form of a hard pan, so hard that neither water nor roots can penetrate below it. The plants are thus shut off from deeper supplies of minerals and the humus-stained water ponds up to the surface in winter flooding, further increasing surface leaching. This is the classic Podsol profile.

The pygmy forest or the leached Podsol, which came first? This is one of those chicken and egg decisions and it is perhaps best simply to conclude that you can't have one without the other. The fact that dwarfing is brought about by lack of nutrients is shown by two natural phenomena and an on-site experiment. Addition of nitrate to pygmy forest plots brings an immediate response in massive growth. Intermixed within the pygmy plots are a few massive Bishop Pines, investigation of which has shown that the main tap root of each one has somehow penetrated the hard pan to the mineral store below. It is also possible to walk a few metres from the most extreme pygmy forest to the flanks of Jug Handle Creek, where you can find soil which, being continuously enriched by erosion from upslope, has maintained its clay content intact, and its exchange surfaces full to overflowing with useful minerals which support magnificent forest.

Hans Jenny, a pioneer of soil science.

Jenny and his collaborators believe—and it is only a belief, they still work to establish final proof—that this ecological staircase as they call it represents in its simplest form a chronosequence, a sequence of events determined in the main by time alone; and that given time, any rock would, under the influence of adequate rain and temperature, go through a similar sequential change: youth, maturity, senescence, enrichment, stability then loss by leaching to a state which could support little in the way of productive growth.

If this is true, why then are not pygmy forests, and thus Podsols more widespread upon the earth? Why don't we find them and their counterparts on every other part of America and, indeed, across the world where rocks have remained exposed to a similar climate for 100,000 years or more?

The anti-staircase lobby as we will call them say that Jenny is wrong; that the pygmy forest is just a localised phenomenon, of nothing but local importance; and invoke many other explanations for its presence: fire; gross climatic change; shifting dunes; and many more. But they don't look at the broad sweep of this enormous chunk of real estate called America.

When considering a subject of such great importance it is best to take a broad perspective. Hans Jenny did just that, and we are going to follow his example and gain some of the experience that is America for ourselves, and in so doing, learn about its soils, their potentials and their limitations.

Chapter Three

The Eastern Connection

THE mountains which form the backbone of the American north-east, Appalachian and south-east farming regions are cloaked in broadleaf deciduous woodlands which are as old as the flower-bearing plants themselves. They can give us an insight, not only into the origin of America's vegetation, but of our foremost garden plants. Even today the remnants of these woodlands can provide a number of surprises, not least the mysterious 'eastern connection' which gave London its Plane tree and in 40 years wiped out one of America's commonest and most useful trees. It also helps to explain the origin of the continent itself.

There are in the world certain combinations of time and place which seem to complement each other in a very special way, each bringing out the best in the other. Spring in the Appalachians is one of the most perfect of these combinations, for it captures the perfection of that which links climate and vegetation. The Appalachians are a great sweep of mountains whose overall orientation approximates to north/south and forms a very elegant backbone to the eastern States: Maine, Vermont, New York, Pennsylvania, the Virginias and Carolinas, Kentucky, Tennessee, Alabama and Georgia. The oldest rocks of which they are formed were laid down more than half a billion years ago and being metamorphosed sedimentary are extremely hard and have stood the test of erosive time. They were folded into mountains around 200 million years ago and it is believed that some have remained at the surface, and hence exposed to the creative power of water, ever since.

If this is indeed the case, and there is no scientific evidence to deny it, they have borne witness to more springtimes than almost any other rocks across the world. The best place I know in which to see these rocks and stand firm on that immense sense of vernal history is in the Great Smokies, which form the heart of the Blue Ridge Mountains. There you would not only enjoy one of the great spectacles of the world, but you would also find that the song is right: above the 1,500-metre elevation the forest is dominated by Fraser Fir and Red Spruce; not a Pine in sight.

In general make-up these upland forests appear to be a southward extension of the coniferous forest or Taiga which dominates much of the Canadian landscape far to the north. Despite the fact that they do have a certain disjointed continuity along the summits of the Appalachians with the great sweep of those northern forests and bear them great resemblance, they are different enough in strict botanic make-up to deserve distinction in their own right. Their correct name is thus Appalachian Sub-Alpine Forest and there were more than one million acres of them before white man came upon the scene.

So the song is right: this is the land of the lonesome pine. But what about Hans Jenny's pygmy-forest theory? Does the Appalachian experience relegate it to the sphere of hillbilly moonshine?

Observation reveals pygmy, or rather dwarf, forest, dominated by Fraser Fir and American Mountain Ash, but all above 1,290 metres where the raw wind whipping up mountain fog even on a summer day warns that the trees which grow hereabouts do so almost on their altitudinal limit. Nothing strange about that; no support for Jenny's theory. We must look again.

One of the problems of walking through a totally forested landscape is that not only can't you see the wood for the trees, but often it is impossible to see the landscape for the same reason. However, from a few of the summits so steep that nothing can grow, it is possible to look down and see areas devoid of large trees scattered throughout the sub-alpine forest zone. These open areas are not all the same; some are covered with grass, some with heathy scrub and some with dwarf trees the vegetation of which includes both genera and species found in the true Pygmy Forests of the far western coast.

Some of these balds, as they are known locally, occupy slopes which are too steep to support taller vegetation, but some are on flatter land, on ridges and mountain crests, and some are underlain by Podsols; these may well represent an overmaturation of the natural process of succession and soil formation similar to that hypothsised by Jenny for the Mendocino coast.

As for the rest, one must admit that the majority of the landscape is covered with full-size forest, and it's all on a slope. The most striking feature of the landscape is that less than 10% of the whole area has a slope of less than 10°—ideal for continued erosion and revitalisation of the soil but not for farming, one reason why the hillbillies who lived in these mountains had to be such a tough lot, and at least part of the reason why the whole area is not covered with pygmy forest. The rocks, and hence the soil, may well have remained exposed to the elements for a long time, but everything else about the Great Smokies indicates that things have not been stable hereabouts, and study shows that the whole area has been subject to massive changes throughout its long history. For example, the balds are full of hillbilly plants like Fiorin and Timothy, and grasses introduced by the farmers in an attempt to improve the grazing have now run wild along with many weeds including Self-heal, Sheep Sorrel, and Hawkweeds.

Signs of Indian encampments are not uncommon and carry back the story of disturbance at least to 6,000 years before the present, when much of North America was a much warmer and drier place than it is today. Then the sub-alpine forests were confined to the coolest, wettest localities on peaks above 2,500 metres in altitude. There they formed islands almost in the sky, each a refugium of this forest type, throughout the hypsithermal—the warm dry period. As the climate grew cooler and wetter, they spread back down on to the flanks of these high peaks to their present lower level of 1,750 metres, the broadleaf deciduous forest being displaced once more to occupy only the lower altitudes. On nearby peaks which do not rise above the critical 2,500 metres, there were no refugia and so today their peaks are swathed to the top with broadleaf forest or are capped by balds. Other explanations can be invoked to explain the present distribution of the sub-alpine forest, but the above seems most adequately to fill the hill.

The warm dry period was not a localised phenomenon. It was felt clear across the northern hemisphere with varying effect, and followed hot on the heels of an opposite and much more massive change. From about 27,000 until 12,500 years ago much of Europe and North America were locked in the grip of an ice age. The ice sheets of America covered an enormous area and their southern limit was a ragged line of destruction across the country from south of the Great Lakes in the east to British Columbia in the west. The effect of the climatic change was, however, felt far to the south. Mountain-top glaciers formed on the peaks of the northern Appalachians and Rockies and the whole growth of vegetation was displaced by the climatic change in terms both of latitude and altitude: the former towards the warmer south, the latter down towards the warmer lowlands. Fortunately the Appalachians run north/south and their massifs are well dissected by a multitude of valleys. They therefore did not form a barrier to southerly migration and more important they provided a multitude of niches at all levels and of all aspects in relation to the warming sun in which relict populations of most species and vegetation types could find a temporary home.

There is evidence that during this time the higher summits of the Smokies reached above the timberline and perhaps it was under these conditions that some of the balds came into being.

The linking of the similar sub-alpine forest of the southern Appalachians and those of the true Taiga by the southwards extension of the latter during the ice age allowed migration of species in both directions. That Jack Pine and Black Spruce, both of which are dominants of the true Taiga, grew in abundance as far south as Georgia and at the feet of the Appalachians is proved by macroscopic remains of readily identifiable wood found in peat and lake deposits. The presence of pollen of Oak, Hornbeam, Beech, Plane, Ash, Walnut, Elm and Hickory amongst the macroscopic remains of the Taiga trees suggests that the broadleaf forest did not disappear from the scene but simply retreated to more favourable situations for the duration of the cold bad times. Especially favoured would have been sheltered situations along the coast of the Atlantic and the Gulf of Mexico, a coastal strip which was then much broader than it is today, for as the ice sheets formed on land withdrawing water from the sea, the level of the latter went down by almost 100 metres.

So the evidence both macro- and microscopic from an ever-increasing scatter of sites across the continent shows that as the ice sheets developed in the north, the broad swathe of coniferous forest was pushed southwards, squeezing the more warmth-demanding broadleaf forest into smaller and smaller refugia. The vegetation and hence the process of soil formation have thus been kept in a continual process of change.

If those refugia had not been there, if the glaciation had been more severe or more prolonged, if the Appalachians had run east/west and formed a barrier to migration, it is a sure fact that the modern Appalachian spring would not be such a riot of flowers. For as the glacial climate grew milder the process was reversed, and over a much shorter period the Taiga drew back to its current northern fringe, the sub-alpine forests to their mountain tops and the broadleaf forest moved north and re-climbed the flanks of the Appalachians, clothing them in all their diverse glory—a glory which is today written anew each spring as the many forest types say it with the flowers of what may well be their two hundred million and oneth spring.

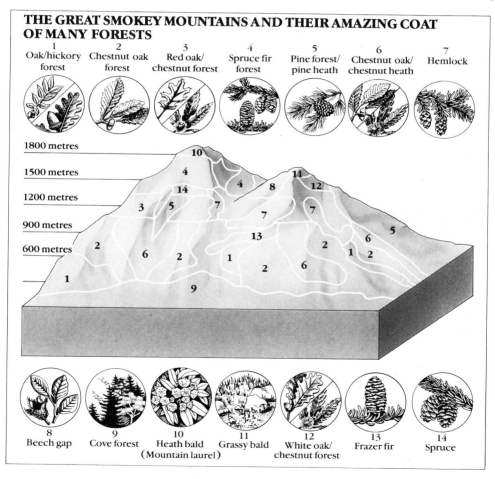

THE GREAT SMOKEY MOUNTAINS AND THEIR AMAZING COAT OF MANY FORESTS

1 Oak/hickory forest
2 Chestnut oak forest
3 Red oak/ chestnut forest
4 Spruce fir forest
5 Pine forest/ pine heath
6 Chestnut oak/ chestnut heath
7 Hemlock

8 Beech gap
9 Cove forest
10 Heath bald (Mountain laurel)
11 Grassy bald
12 White oak/ chestnut forest
13 Frazer fir
14 Spruce

The Cove hardwood forests are doubtless amongst the most beautiful deciduous forests in the world. They occupy the rich soils of the coves at the feet of the mountains where the soils are continuously enriched by downwash from above. Six tree species, their average height rising up to between thirty and forty-five metres, share dominance: Eastern Hemlock, which probably arrived from the north during the last glaciation; Yellow Buckeye; Silver-bells; White Basswood; Sugar Maple; and Yellow Birch. Tulip Tree and American Beech are important in some stands. Together these eight species make up 80 to 90% of the canopy, but mixed in with them are 32 other species, and the whole diverse canopy provides shade for many hundreds of species of plant. It is the most marvellous place and almost impossible to describe.

Where can I start? For the whole area from the lowest deepest cove to the highest bald is a botanist's paradise. All I can do is suggest that you make the pilgrimage and see it for yourself. I can assure you that you won't be alone, for on the last Thursday, Friday, and Saturday in April the tourists come in their tens of thousands and again through mid-June to mid-July. Everywhere you look there are flowers, growing singly or massed in all their glory. You may take your choice how you view them but do not pick; leave this natural beauty for others to enjoy.

Here are just a few to whet your appetites.

White-fringed Phacelia is in places so abundant that it looks as if snow has covered the forest floor in May. The true beauty of its fringed flower is only seen in close-up, when it looks as if it has been fashioned out of finest porcelain.

Purple Rhododendron, often mixed with Mountain Laurel, picks out the heath balds or laurel slicks as some locals call them, an unforgettable palate of purple and pink, while White Rhododendron livens the more sombre depths of moist stream sides at all elevations.

Sassafras, a half-canopy tree growing to thirty metres, produces great sprays of yellow flowers, some male, some female. Its roots, twigs and bark yield aromatic Oil of Sassafras from which a spring-tonic tea is made.

Silver-bells dangle in May in elegant sprays from a medium-sized tree of that name which is abundant in rich loamy soils where its roots can penetrate to a great depth.

Witchhazel flowers once its leaves have gone, its bare twigs covered in yellow spider-like blooms from October through to January. It is thus both the last and first plant to flower in the forest and provides a fragrance which I will always associate with grazed childhood knees, for an astringent compounded from its petals was a cure always used by my granny.

No fewer than three Magnolias grace the slopes of the Smokies and all are fairly common up to the 1,300 metre mark. Their cream-white flowers which appear in April and May have petals up to twelve inches in length and are thus the largest, but unfortunately are among the most evanescent, in the forest.

Tulip Tree, a close relative of the Magnolias, can soar up to 60 metres in height and such a specimen when covered with its cream-yellow flowers and giant four-lobed leaves is enough to take anyone's breath away.

I could have chosen others—to be exact, more than 1,290 others—but these must suffice. The first three, because they occur in such abundance that they become the landscape and at the right time of year you must look hard between the Rhododendrons to see the trees. In Britain we are used to seeing Rhododendrons growing as 'weeds' out of their natural place. They were introduced to Britain from America and Asia to grace our English gardens and to produce cover in the estates which surround our stately homes. They are not my favourite plant, for when neglected, they so often run amok in a riot of woody destruction. Here in the Appalachians they are at home and having a specific role to play within the diversity of the forest, know their place and are kept within it through the interplay of opportunity and competition which is the society of plants.

It is of more than passing interest that the majority of our most showy and highly prized Rhododendrons and Azaleas (a name given by horticulture to the deciduous forms) are the result of hybridisation of American and Asian stock.

The first experimental plant hybrid reported in the literature was of a cross made in 1717 between two Pinks by one Thomas Fairchild, a commercial flower-grower of London. Much earlier than this another hybridisation had perhaps taken place in London, although this was accidental. The resultant plant was, however, going to grace the streets of that great city and many others across the world almost from that day forth. The accident, if such it was, may well have taken place in the gardens of a house in Lambeth, in a garden owned and run as a nursery by the John Tradescants, father and son. The Old World Plane (*Platanus orientalis*) was growing in their garden before 1633 and the New World Plane (*Platanus occidentalis*) was added to the collection later. There is little doubt that both were grown if not actually introduced into Britain by these men, for both were plant hunters and on at least three occasions John the son made collecting trips to Virginia, where the New World species grows.

Oriental Plane (*left*) and London Plane, an accidental hybrid.

The first hybrid tree now known as *Platanus* × *acerifolia* was noted in the Botanic Garden at Oxford in 1670. John the elder had been appointed keeper there in 1637, although he was then too ill to take up the honour. As Mea Allen suggests in her excellent book on the Tradescants, 'When the new Oxford Botanic Garden was being stocked, botanists sent gifts of plants and trees. It would be strange if John the Younger did not send some saplings or plants in memory of his father.' Perhaps, she surmises, amongst them was the accidental hybrid: the first London

Plane, the graceful tree so popular now in towns, for it has the pleasant habit of cleansing its trunk of soot by scaling off its bark in patches. This gives it the appearance even on the dullest day of being dappled with sunshine.

Even if this surmise is not true, we have the Tradescants to thank for many other introductions from eastern North America which, either as hybrids or in pure form, have become an important part of every English garden: Shirley Poppies, Ten-week Stocks, Perennial Lupins, Michaelmas Daisies, Herbaceous Phloxes, Golden-Rod, Stags' Horn Sumach, Dogwood, Doronicum, Passion Flower, Virginia Creeper, Scarlet Runner Bean, and many more. What would the living space of our homes be like without window to window fitted Spider plant? (*Tradescantia virginiana*, named after the men who saw its potential and the state from which it was first collected.)

It was a great thrill to stand in the heart of the Appalachians under the shade of a *Platanus occidentalis* which they call Sycamore there, and see growing alongside, Creeping Phlox, Bleeding Heart, two plants now considered to be typical of the English garden, and Dandelion and Plantain, two English weeds that came the other way.

However, if you want to see weeds in all their invading glory you don't have to look far in Appalachia. Two woody vines, Japanese Honeysuckle and Kudzu, festoon every vantage point along many roads and tracks in the area. Introduced from eastern Asia, they found what is for them unoccupied space and they have run amok, threatening the forest societies which are the heart of the Appalachians, for if something isn't done to check this spread, certain forest areas could change out of all recognition—nature aided by man, now out of man's control.

There is little doubt that the most tragic introduction from the Old World was that of *Endothia parasitica*, a parasitic fungus. There is some indication, but no sound evidence, that it arrived around 1895 on young chestnut trees brought from China or Japan to be planted in New York's Zoological Garden. Back home in eastern Asia it caused a blight disease of the local species of chestnut which perhaps, due to long exposure, had built up some resistance and immunity. Not so in America. The American Chestnut was highly susceptible and soon succumbed. By 1904, chestnuts in the vicinity of the Bronx Zoo were infected and were dying. The microscopic threads of the fungus grows throughout the tree, sapping its strength and killing living cells. Only later does the fungus begin to protrude through the bark. Mere pimples at first, the fruiting bodies then burst into nasty-looking yellowish pustules which in a few days enlarge into curved horns. Each horn is capable after rain of delivering five million spores, each capable of infecting another tree and starting the whole process off once more. Dying trees acted as centres of dispersion of the disease. One woodpecker was found carrying three-quarters of a million spores on its feathers. Hollow trees with a hole both at base and near the crown acted like factory chimneys, sucking air in at the bottom and sending a plume of deadly spores high into the breeze.

By 1925 it was realised that the blight was on the rampage and by the early 1940s, all the trees were dead. It is so easy to say, but to comprehend the enormity of this destructive change is no easy matter. Take into account that four of the main forest types which covered the Appalachians and their hinterland from Maine clear down to Georgia were in 1925 co-dominated by *Castanea dentata* (the American Chestnut) which made up between 30 and 70% of each forest stand. Twenty years later there were, in effect, none left. The fact that a seemingly dead tree will re-sprout from round its base may seem a good omen at first, but these new shoots are infected and soon succumb again. So devastating was the disease that a book on the important trees of the Appalachians published by the Forest Service in 1970 does not mention the Chestnut. So twenty years brought about massive changes in the plant society of these forests, opening up opportunities for other trees like Oak and changing the pattern of light and litter reaching the forest floor, thus effecting the make up of the ground flora and fauna: a massive change in the natural order of things, but much more than that, a catastrophe for man.

In early summer flower the Chestnut added its own beauty to the forests and because they flowered after the last frost they always set an abundance of fruit for animals and man alike. Appalachian bears gorged themselves into hibernation on their rich white 'meat', which when boiled or roasted was an autumn staple, both for the Indians and later for the Pioneers. The chestnut's decay-resistant wood proved ideal for log cabins, shingles, fence posts and rails, cross-ties on rail roads and telegraph poles. Its worked beauty made it ideal for panelling and furniture and what was left or was produced by coppice management provided pulp for paper mills and fuel for brick kilns and smelting furnaces. Furthermore, its wood and bark were rich in tannin, essential to the leather industry, then amongst the most important. All this great resource was wiped out by the introduction of one microscopic parasite from the east of Asia.

The link between the Appalachians and Eastern Asia is not all bad, however, as we shall see, but how did this eastern connection come about?

Apart from their beauty, the other eight plants which I selected from the Appalachian list show another truly amazing feature in common. They are amongst the world's best examples of disjunction, that is, of a plant type which has its natural home in two or more widely separated areas of the world and nowhere in between. Sassafras, Noah's Ark, Silver-bells, Witchhazel, Magnolia and Tulip Tree are members of six genera whose species are found only in eastern North America and the far east of Asia. *Sassafras* has one species in the former and two in the latter; *Cladrastris* 1 and 4; *Panax* 2 and 3; *Hamamelis* 3 and 6; *Magnolia* 8 and 30; and *Liriodendron* 1 and 1. The botanical exploration of the Appalachians was started in earnest by a local-born farmer, one John Bartram, 1699–1777, whose fame was established when his name was given to the Little Apple Moss, *Bartramia*. The work, which is still being carried on today by Prof. Jack Sharp of the Big Orange University at Knoxville, Tennessee, has confirmed the evidence of the existence of this Eastern Connection.

Today, thanks to their expertise and dedication and that of many other botanists, we know that at least 56 genera of flowering plants demonstrate the disjunction linking these two far-flung areas of earth. We also know that the majority of them are woody, deciduous with simple undivided leaves and seeds and fruits which bear no special means of dispersal. All the four hundred species in the group are perennials with heavy rootstocks and are thus adapted to life on the floor of a deciduous forest. They bring forth their leaves in spring, charge their energy banks while there is light and then disappear underground later in the year when the dense shade cast by the canopy trees above makes photosynthesis impossible.

The link could therefore be purely climatic: two areas with the same temperate climate. If this is so, what of the temperate areas in between which lack these plants? And what of the other temperate plants shared by these two areas which are also widespread in other parts of the world?

It was Asa Gray, 1810–1888, one-time director of the Botanic Garden in Boston, contemporary of, and correspondent with, Charles Darwin and Sir William Hooker, the latter of the Royal Botanic Gardens at Kew, who first pointed out these disjunctions and posed the question, how did such a disparity come into being?

Apart from their shared plants, these two areas have another thing in common. The rocks of which certain of their mountains are composed are very old and some have been exposed—yes, you have guessed— for the same two hundred million years of geologic time.

We now have much more than reason to believe that at that seemingly dim and distant time, these two areas were much closer. In fact, both were parts of one super-continent, Laurasia, which was just about to split off from Gondwana-land, the other more southern half of a super-super-continent, Pangaea.

Much of the central part of the greatest mass of land which has ever existed here on earth must have been characterised by deserts, for it is unlikely that they would have ever felt the benefit of rain-bearing winds. Closer to the margins of the land mass there were tracts of well-watered land and enormous coastal swamps which stretched from tropic into temperate latitude, across the vast bulk of land set astride the equator. These coastal swamps were inhabited by amphibians, reptiles and the first mammals, all of which lived on an abundance of plant material: giant club mosses, horsetails, ferns and cone-bearing plants in great variety. There were, as yet, no flowering plants to grace the newly forming soils, no birds to fly amongst their leaves and the dinosaurs were only then beginning their tyrant reign.

We know all this from the fossil record, both in the rocks of the Grand Canyon Staircase, and many other less spectacular sites and sights across the world.

At first the paleontologists, that is the up-market 'rock hounds' who specialise in fossils, concentrated their attention on the more spectacular remains, wood, skulls, cones and the like. Though abundant in places, this part of the fossil evidence has been handed down to us by mere chance, the chance that the animal or plant to which in life they had belonged, died in a place in which their remains would undergo the process of fossilisation rather than that of decomposition. It is, therefore, most likely that the macro-fossil record must perforce be less than complete and there will thus always be links missing in our chain of knowledge.

With plants there is an added problem, and yet an added advantage. Many of the plants of the past of which we would like to know much more grew in drier places, far beyond the compass of coastal swamps, and unlike animals, they could not walk to their death in such stations of fossilisation. However, being often far removed from any permanent body of water the true land plants, the cone- and flower-bearers, evolved a mechanism by which the male gametes, or at least the genetic information they contained, was packaged and transported to the female, where fertilisation took place. The vehicle of transportation is the pollen grain and in the main, the transport is by air. In order to overcome the enormous problems relating to the success of such a system, pollen grains are produced in their tens of millions and each genetic message is fresh wrapped in Sporopollenin, a chemical so resistant to decay, and hence ripe for fossilisation, that it has been found in some of the oldest sedimentary rocks in the world. It is also of great interest that it is found in the spores of many of the more lowly plants, algae, fungi and mosses, the ferns and their allies and that in the pollen grains of the cone- and flower-bearers, it is built into a complex wall structure which in many cases may be used to identify the plant down to group, family, genus, even in some cases to specific level. Thus it is that ever since cone-bearers have born cones and the flower-bearers flowers, there has been a constant rain of potential microfossils across the continents of that dry land which at the time of their evolution were one, Pangaea.

The first real macrofossil of a flowering plant we have on record is from Jurassic (180-million-year-old) rocks in Russia. It is a fruit with small seeds each of which bears a parachute-like pappus: its given name, *Problematospermum*, indicates the difficulty in assigning it a true identity.

Sculptured monosulcate pollen undoubtedly produced by angiosperms is known from 126 million years ago in rocks from England, Maryland and Argentina and 10 million years later there was an abundance of such pollen types blowing across the world.

So it would appear that some time as the supercontinent split and more of its area was opened up to the effect of rain-bearing winds the flower-bearing plants had their origin, and taking advantage of all that new unoccupied well-watered space on offer, exploded across the drifting continents in a variety of beauty.

Current evidence suggests that they originated somewhere in the temperate subtropical interface of what was West Gondwanaland. There is also little doubt that the first flower-bearers were woody plants, perhaps akin to the contemporary order *Annonales*, within whose taxonomic compass lies the family *Magnoliaceae*, which includes our three Magnolias and the Tulip Tree, all of disjunct fame.

Could it be that they and the others of their disjunct ilk are all that remain of an ancient flora which occupied the temperate areas of Pangaea all that time ago? Perhaps an enormous forest covered much of what is now America, Europe and Asia at some time in the past and has only survived in those two areas the rocks of which have remained exposed to a climatic regime that has always favoured temperate deciduous woodland.

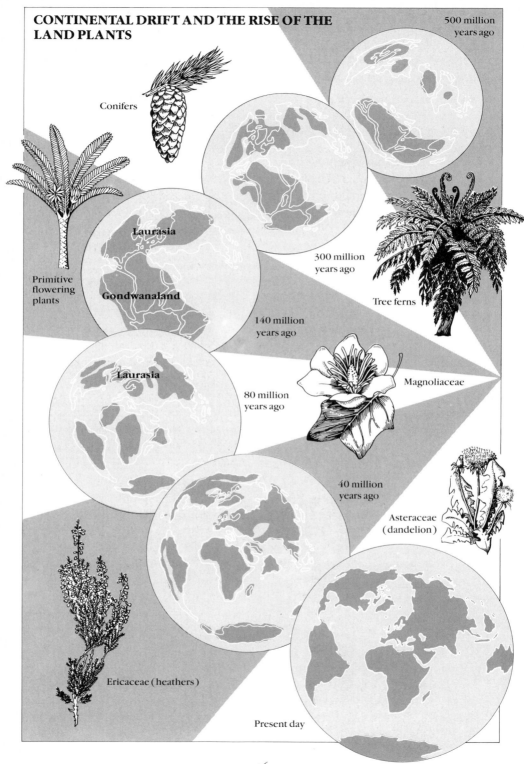

CONTINENTAL DRIFT AND THE RISE OF THE LAND PLANTS

500 million years ago

Conifers

300 million years ago

Laurasia

Gondwanaland

Primitive flowering plants

Tree ferns

140 million years ago

Laurasia

80 million years ago

Magnoliaceae

40 million years ago

Asteraceae (dandelion)

Ericaceae (heathers)

Present day

As more concerning continental drift became known the fact that this was the case became more clear. But what of the other parts of Europe and Asia which have remained in temperate latitudes throughout the relevant time? Why has this ancient enclave of plants disappeared from their good land? The ice age offers an answer. The broad sweep of Eurasia is so great that its central plains, though temperate in latitude, have the climate of semi-desert and thus support a specialised flora. Perhaps more important, the main mountain chains within the temperate belt run east/west except those in eastern Asia, which, like the Appalachians, run north/south.

During the four glaciations of the last one million years, the east/west mountain chains would have formed a barrier to the southward migration of plants and animals. The ice developing both in the north and on the tops of the mountains would thus have brought about massive extinctions as glacial conditions spread within its aegis. As we have already seen, the predominantly north/south orientation of the mountains in North America and eastern Asia did not; in fact, they offered a highroad of varying elevation, a diversity of habitat for migration in either direction.

Much earlier, some 60 million years ago, as North America rotated anti-clockwise and northwards away from Europe, the main diversity of the flowering plants had evolved and so species and genera were able to migrate the long route round, across what is now Scandinavia, Iceland, Greenland, the Canadian Arctic islands into the bounty of the New World. However, only those which could tolerate the long cold winters of the high latitudes could make it across the sub-polar route, and so the two regufia of temperate forest came into existence, the enormous area between them devoid of their diverse presence.

It is little wonder that we in Europe had to re-import so many plants back from North America and eastern Asia to grace our hybrid gardens.

So if you do decide to make the Appalachian Pilgrimage you can make contact, with the antiquity not only of temperate forest, but of the flowering plants themselves, rooted in soils which have since their first formation been subject to changes of many types and many magnitudes, all of which have kept them rejuvenated and in good heart. This provides a maturity of sylvan experience which can only be found in one other part of the world and which, when the pioneers first came upon the Appalachian Spring, stretched west to the Mississippi and beyond.

CHAPTER FOUR

Rooted in the Past

THE northern and southern plains are a vast area of predominantly flat land which separates the well-watered woodlands of the east from the high dry deserts of the Rockies to the west: a vast tract of Mollisols bearing grassland (prairie to the locals) which once supported herds of game as big and as diverse as those which sadly are found only in Africa today. This is the battleground of the Great American Interchange, where the flora and fauna of two long-separated continents met on the formation of the Isthmus of Panama only 3.5 million years ago. It is a story of mass extinction, in which man may well have played a part.

West of the Mississippi, the environment begins to change as your journey takes you away from the life-giving sea which every day breathes oxygen-rich vapours which form clouds shading the dry earth from the full evaporative power of the sun and bringing the promise of rain. The balance between precipitation and evaporation gradually shifts in favour of the latter and the supply of water both for the growth of vegetation and for the process of soil formation both diminishes and becomes more seasonal. Water held in the soil and thus available to plant growth is best regarded as 'hanging' water. Where rainfall always exceeds evaporation, the hanging space becomes over-full and excess drains away down through the soil, carrying minerals and nutrients with it; the end result being a leached acidic Ultisol, or in its extreme form, a Podsol.

Where evaporation reigns supreme, the overall movement of water through the developing soil will be in the other direction, evaporation drawing water back up towards the surface. The end result of this pedogenic (soil-forming) process are the Aridisols of deserts and semi-deserts. Such soils became enriched with minerals, so much so that they are saline and/or alkaline, their lower layers saturated with the less soluble minerals like chalk and gypsum.

Intermediate between these extemes of leaching and enrichment are the Mollisols. Formed in an environment in which there is a year-long struggle for supremacy between leaching and evaporation, the soils are thus rich, but not too rich, in minerals and nutrients, their lower layers often marked by nodules of chalk, calcium carbonate. These soft soils are usually found in relation to grasslands or forests, the floors of which are rich in ferns and mosses.

Moving down from the Appalachians towards the Mississippi and beyond, the lush continuity of forest begins to change, at first in a subtle way—a slight shift in the range of species up or down slope, each sticking as it were to its optimum soil-water regime; a loss in stature of the dominant trees; an opening up of the tree canopy and a closing of the shrub and herb layers; Oak/Hickory and then Oak alone assuming dominance. New light-demanding species make their appearance on the forest floor and for the first time natural open glades, no longer dotted with woody shrubs and saplings, but dominated by the massed presence of graminoids, gradually become more and more the order of your westward way.

Graminoids are plants which look like, feel like, and above all possess the properties of, grass, properties which allow them to survive, indeed to thrive on, being grazed.

There are in the world some 10,000 sorts of grass and 3,000 sorts of sedge, plants whose vegetative bulk is made up not of stems, but of long narrow leaves which put on new growth at their base rather than at their apices. These are the true graminoids. Add to them all the perennial herbs which bear their buds and hence their growing points at ground level and you include the successful world of the Hemicryptophytes.

Their success relies on the fact that however interesting and important soil may be to man, grazing animals do not like to eat it. So instead of gritting their teeth at each mouthful they set them somewhere just above soil level and thus leave the buds and the basal growing points behind, to carry on their all-important work of regeneration and the production of another crop of leaves.

So it is that west of the Mississippi, forest gradually gives way to grasslands, or as they are called in the mid-west, prairie. I first came into intimate contact

The rolling landscape of the tall grass prairie in Kansas.

with the forest/prairie boundary while on a trip by road west from Tulsa to the small town of Pawhuska. The road wound (well, wound by American freeway standards) through a rolling landscape decked with clapboard farmsteads and well wooded, with Walnut, Elm, Hickory and Sycamore prominent, though Oak was always dominant and Pecan abundant along the river flats. Arriving at Pawhuska, I was somewhat worried because I had made the trip in order to see the bright golden haze of thousands of square kilometres of tall-grass prairie, and so far I had seen none.

Checking in at the motel, I awaited our contact, one Dick Whetsell.

A knock on the door announced his presence, a presence, I am sure he won't mind me saying, straight off the set of *Oklahoma*: cowboy boots, blue denim, leather belt with lazy-A buckle, and a stetson covering a shock of silver hair and an immensity of knowledge about Big Bluestem and the many other grasses which go to make up the renewable wealth of this very diverse slab of real estate. The biggest 'meader' in the world, all 120 million hectares of it.

Soon we were on our way, off the paved highway and on to the dirt road that took us to the edge of the real prairie.

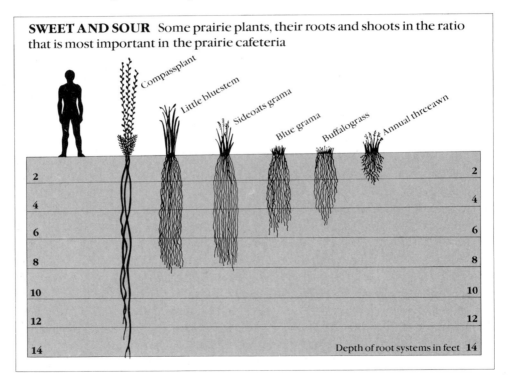

SWEET AND SOUR Some prairie plants, their roots and shoots in the ratio that is most important in the prairie cafeteria

Compassplant
Little bluestem
Sideoats grama
Blue grama
Buffalograss
Annual threeawn

2   2
4   4
6   6
8   8
10   10
12   12
14   Depth of root systems in feet 14

I was, and always will continue to be, amazed not only by the immensity of the prairie itself, but by the story it tells, in beautiful and intimate detail.

First, its diversity. From a distance the tall-grass prairie, for that is what I stood within, looks a more or less uniform mass, an untidy lawn of gigantic scale, for at least three of the dominant grasses stood as tall or taller than my 186 cms. A close look, however, reveals an unending diversity of size, pattern, shape and species. The dominant components are without doubt the true grasses, and across the length and breadth of the prairie there are at least 80 species. With names like Arrowfeather Threeawn, Big Bluestem, Curlymesquite, Grama Sideoats, Kentucky Bluegrass, Nimblewill, Tumblegrass, Vine Mesquite and Weeping Lovegrass to choose from you can almost smell the mixture of hay, woodsmoke and saddlesoap. These are backed up by a great diversity of legumes: members of the Pea Flower family; composites; members of the Daisy Flower family; herbs, forbs in American parlance, members of a whole host of other families, each evolved to play a part in life on the prairies.

The members of each group range in size from monsters like Big Bluestem (a grass), Lespedeza (a legume), Compassplant (a composite) and Plains Larkspur (a forb), each of which grow to around 2 metres in height, to diminutive Buffalo-grass, Prairie Acacia, Blacksampson, and Woolly Plantain, all of which rarely top 15 cms. The vast majority of the plants are perennials, their roots and stems forming mats or bunches which help to bind the soft soils and protect them from the erosive power of the heavy raindrops of prairie storms. Annuals, however, can be abundant at times, filling in the spaces and seasons with their passing presence.

A good rule of thumb, especially when dealing with the grasses, is the taller and more tussocky the top, the longer and more bulky will be the roots. This is not entirely true—for example, the Dotted Gay Feather's roots may reach down to as much as 5 metres, whereas its shoots are rarely more than 70 cms tall—but in the main a tall shoot means a deep root. This shoot-to-shoot relationship focuses our attention on that part of the plant which, since we usually do not see it, we often don't think too much about.

Roots reach down into the soil in order to exploit its hanging water and the store of nutrients it holds in available form. In so doing, the roots also anchor the plant and bind the soil together. Plants are frugal creatures and so do not waste energy in sending roots down into soil in which there is no water. Hence it is another rule of thumb that the depth of rooting corresponds to the depth of the hanging-water reservoir, the lower parts of which are on the prairie often marked out by an abundance of nodules of calcium carbonate. Roots will only penetrate deeper than this if there is some underground supply of water which is unrelated to the local balance between rainfall and evaporation.

So the height of the prairie grass tells you a lot about the prairie environment. Moving west of Pawhuska, the Tall-grass prairie is dominated by Big Bluestem, Switch and Indian Grass. Further west, much further in Colorado and beyond, is Short-grass Prairie dominated by the Grama and Buffalo grasses, reaching only as high as a Prairie Dog's eye. This includes many semi-desert plants, including the famous Tumbleweed and infamous Prickly Pears. Between these extremes is a broad swathe of Mixed-grass Prairie which is just that, a mixture of all types not as tall as the tall grass, nor as short as the short grass.

It is of great interest that overgrazing tends to have much the same effect as increasing aridity, in that it causes a scaling down in the structure of the whole grassland complex.

The growth of the shoots depends on adequate water supply and hence an optimum growth and penetration of the root. Likewise, good root growth depends upon adequate green shoot material to speed the products of photosynthesis for growth and storage in the roots—a case of you can't have one without the other. Grazing must reduce the photosynthetic potential of the shoots and hence limit root growth. Overgrazing can therefore leave a grass plant with a stunted and weakened root system at the end of the year. Such a plant will be able to respond only poorly to the rustle of the next spring rain and hence there will be a knock-on effect which may further stunt its growth. Continued overgrazing will, in time, change the composition of the prairie towards domination by less palatable and even non-edible and poisonous plants.

Herein lies the true magic of the natural prairies, a diversity of plants of different stature, the roots of which reach down to tap the whole of the hanging reservoir of water. Many of the taller, deeper-rooting forms had their origins in the warmer south and so flower only in the later summer, drawing on the deepest reserves of soil water. The lesser grasses like Grama and Buffalograss, flower earlier in the year and thus complete their life cycle before the surface layers of the soil in which they root, dry out. Annuals such as Witchgrass and the Dropseeds are able to nip in, as it were, when the opportunity arises; they germinate, leaf, flower and fruit all in one short season and then drop their seeds to lie dormant in the soil. The whole fruiting head of the Witchgrass breaks off and rolls about like Tumbleweed, the wind speeding it on its way, thus aiding its dispersal.

What is more, and perhaps most, important, the grass plants themselves respond to drought by changing their metabolism, that is their own inner chemistry. One effect of this change is that they become less palatable and so as the season progresses, the grazing animals move on to the deeper-rooting, still 'sweet' forms. Under natural conditions this spreads the grazing pressure across the different species throughout the year and so ensures that the plants are never overgrazed. The whole mixed sward thus maintains its optimum shoot/root ratio from year to year and the prairie remains productive.

In this way everything in the prairie garden remains lovely and there is no lovelier sight than the prairie in June, when the Big Bluestem is as tall as an elephant's eye and more than 70 different plants are in full flower.

Despite the overall diversity of the prairie plants, they can be conveniently grouped into two main categories, functional groups, best called 'cool season' and 'warm season'.

The former start their growing early in the year and make their best growth when the weather is relatively cool. The surface layers of the soil wetted by spring rains will warm first and help the growth of these shallow-rooted plants. Only later in the season will the deeper layers of soil begin to warm through, turning the warm-season plants on to full growth.

The difference between these two groups of plants goes much deeper than their roots. Each have evolved with their own specific internal biochemistry. The cool season plants are, in the main, C_3 or Calvin plants. Calvin is an American plant biochemist who was the first to work out the intricacies of the chemistry of photosynthesis and he gave his name to this basic biochemical pathway, the first energy-rich product of which is a compound containing three carbon atoms; hence C_3.

The warm season plants are C_4 or Hatch Slack, for they exhibit a different pathway of photosynthesis which produces a four-carbon sugar by a mechanism which was elucidated by two Australian biochemists by the name of Hatch and Slack.

Suffice it to say that both store the sun's energy in the form of sugar, but the Hatch Slack pathway is the more efficient of the two. What is more, the whole shape and anatomy of these plants is structured to maximum efficiency, thus allowing them to operate flat out once the conditions for photosynthesis are right. Such conditions, of course, include an adequate supply of water, for water is one of the raw materials of the whole process.

This was it. At last the 'prairie experience'. I was thrilled. I had read a great deal about these highly integrated plant communities, but it was fantastic to see it all for myself, especially under the guidance of an expert.

Late that golden afternoon, I had just clocked up my 58th—or was it 59th?—flower and was lying down in a patch of Big Bluestem, trying to get a photograph of it against the sun. I panned my camera left and there was Big Bluestem himself, blue denims, hat off, silver hair streaming in the afternoon breeze.

It was a fantastic day, the first of many, and it was rounded off by a superb meal at the local golf club. Conversation centred on grass and the practical questions of range management: the need for regular controlled burning to remove the mass of litter which, uneaten by the cattle, accumulates each year, a useless store of energy and minerals and posing a real danger, if allowed to accumulate for too many years, of a massive fire the heat of which would reach down into the soil and kill the living buds on which the regeneration of the next year's crop so depended. I was amazed by the fact that despite regular burning by man in 1965 there had that year nevertheless been one natural grass fire to every 5,000 hectares of prairie: nature's own management still going its own fiery way.

It was very late as we drove back to the motel and as we did so we saw two animals making good their escape from the glare of our headlights. One was a Raccoon, a placental mammal, the other an Opossum, a marsupial.

Everyone knows that Marsupials are pouched mammals—Kangaroos, Koalas, and the like—and that they inhabit Australia where theory has it that they were isolated by continental drift before the hurly-burly of mainstream mammalian evolution produced the great diversity of placental mammals. To prove the point, and despite the fact that there are a number of placental mammals, including man, which got to Australia under their own steam, the bulk of the roles usually filled by mammals are on the Australian scene filled by marsupials: pouched grazers, browsers, fruit eaters, tree climbers, omnivores, insect eaters and carnivores.

The history of South America from the point of view of continental drift is much the same as that of Australia. This great land mass started its life as part of the giant super-continent Pangaea from which it parted company to become an island some 65 million years ago. It then floated away into the limbo of isolation and so remained until some 3.6 million years ago. During this lengthy period, like a giant Noah's Ark, it rafted its own complement of evolutionary effort west by north across the globe, leaving the South Atlantic ocean and the mainstream of mammalian evolution in its wake. Marsupials thus came to fill many mammalian roles in South America, only the main herbivores being placental.

In contrast, North America, which maintained at least a stepping-stone connection with Eurasia, sailed into modern times devoid of marsupials, with all the mammal niches filled by the more advanced placentals. With all this time to play with, chance, or as it is best called, waif-dispersal, must have occurred. If Thor Heyerdahl could make a successful crossing of a full-scale modern ocean on a raft made out of Balsa wood, so too it may be argued that natural rafts of other trees could carry both plants and animals along the same or other routes.

The fossil record makes it clear that round about 38 million years ago a primate and a rodent made the crossing from Africa to South America, enriching its fauna. The likelihood of a successful crossing in the other direction by a marsupial is

DOWNTOWN LOS ANGELES ONLY 20,000 YEARS AGO

| Dire Wolf | Imperial Mammoth | Western Horse | Bison |

more limited, for the simple reason that even if it did make a landfall, it would have been faced with a landscape full of more advanced placentals. Another problem of waif dispersal, whatever the species dispersed, lies in the fact that only a few individuals would be transported at a time and so the genetic diversity of the pioneer population would be very small.

However, around 6 million years ago great changes started to take place in the fauna of South America, most important amongst which were a rapid rise in the number of placental mammals. The reason was that South America was drifting closer and closer to North America, closing the Bolivar Trough, and to prove it, both advanced rodents and raccoon-like animals arrived in force.

3.5 million years ago, the two island continents became joined by the Panamanian land bridge and the Great American Interchange began in earnest. Animals moved in both directions, not two by two, but in herds, flocks, prides and troops, exploding north and south. The glaciations that were to be the main features of the future years not only widened the developing land bridge, but they also had the effect of clearing away much of the forest which had covered the wetter tropical latitudes and acted as a barrier, especially to migration by dwellers of the more open plains. Glaciations not only lock up water, thereby lowering the level of the sea, they

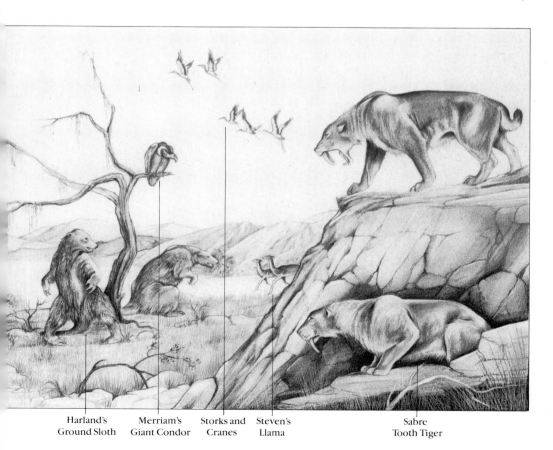

| Harland's Ground Sloth | Merriam's Giant Condor | Storks and Cranes | Steven's Llama | Sabre Tooth Tiger |

also have a radical effect on the earth's climate. One such effect was to aridify the tropical latitudes of America; so much so that vast tracts which now support lush rain forest could at best support a matrix of riverside forest interspersed with open grassland.

During this critical period of continental interchange potential grazing thus stretched from the margin of the ice in Southern Canada down to and beyond 40° South: a mosaic of forest and grassland, freeways of migration for many plants and animals.

The detail of the Great American Interchange would fill many books with fascinating stories of competition for limited resources, of battles won and lost, and of countless animals going to the wall of extinction. In broadest detail the fossil evidence shows that before the interchange, North America was inhabited by 25 families of land mammals compared to the 30 on the southern continent. What is more, no families were common to the faunas of both land masses. During the interchange period, the number of families represented in each area increased dramatically. The drama centred on the fact that resources were not unlimited and not every thing could survive. Interchange over, the number of families soon, at least in evolutionary terms, settled back almost to their original number, the main difference being that they were now a mixture of the two original faunas. It is also true to say that in the fracas the families of the north, on average, fared better than the families of the south, and the marsupial fared worst of all.

Only one member of the marsupial family has survived in North America until today and that is the Opossum. The reason for its unique and signal success lies in the fact that it is a very efficient omnivore and has therefore been able to co-exist with its rival placental competitors like the Raccoon.

An omnivorous opossum associates with Man to its own advantage.

Further work on the fossils already found and the discovery of others still waiting in the ground may well throw new light on the details of the Great Interchange, but it seems unlikely that members of many other families will appear, to complicate the picture and cloud these conclusions. In broad terms, it seems safe to say that 25 or 30 mammal families represent saturation of the available niches and resources present in North and South America respectively. Not even a major upheaval like the mixing of two long-isolated faunas can shake that as a foundation of ecological fact. The carrying capacity, potential if you like, of any landscape, however large, diverse and well endowed it may seem, is limited and that limit cannot be overstepped, at least in the long term.

The evidence for this the Great American Interchange and the conclusion of limitation which it forces upon thinking man lies in a scatter of sites, some fossil-bearing, but others, of much more recent origin, containing organic remains— bones, skin, fur, teeth and dung. Without doubt, the best site in which to take stock of the results of these momentous happenings and the past glories of the megafauna of America is in downtown Los Angeles. No, not Disneyland but in what used to be Rancho La Brea. Set amongst the high-rise buildings of the insurance belt you will find a 26-acre lot, comprising parkland set around a sticky pool of gently bubbling asphalt. Beside the pool, a male Imperial Mammoth stands distraught beside a calf watching his helpless mate, trapped in the cloying asphalt: life-sized models made of glassfibre.

That such a scene took place on that spot many times is known from the remains of these and many other animals which have been found in excavations from other tar pits round about. The fact that no remains have been found in the deep open lakes that are in the main associated with hard asphalt-impregnated sand and clay mixed with gravel and other waterborne deposits, and that many of the bones often contain the remains of blowfly larvae and carrion beetles showing that they were exposed on the surface long enough to rot before being buried and preserved, point to the following type of ancient tragedy. A large animal comes to drink water which has collected on top of the tarry sands; his or her feet get stuck so that the animal cannot escape from the hungry predators and scavengers which soon gather on the scene; some of these likewise become enmired; much later, their remains, probably picked clean of much of the untainted flesh, become covered by sand and gravel washed in by torrential rain. What is more, the record is so clear that we know it happened again and again, and there are always abundant remains of carnivores, both mammals and birds, to lend weight to the supposition.

Although the site had been recognised and documented since 1769, it wasn't until 1905 that scientific investigations were begun and whenever funding was available have continued to this day, revealing a most amazing story and a fauna which, if not for the scientific records, would be thought straight out of science fiction.

The record of the site covers the last 40,000 years; that is the period of the last glaciation and the current interglacial, if such it is. The fauna in this post interchange period include the following:

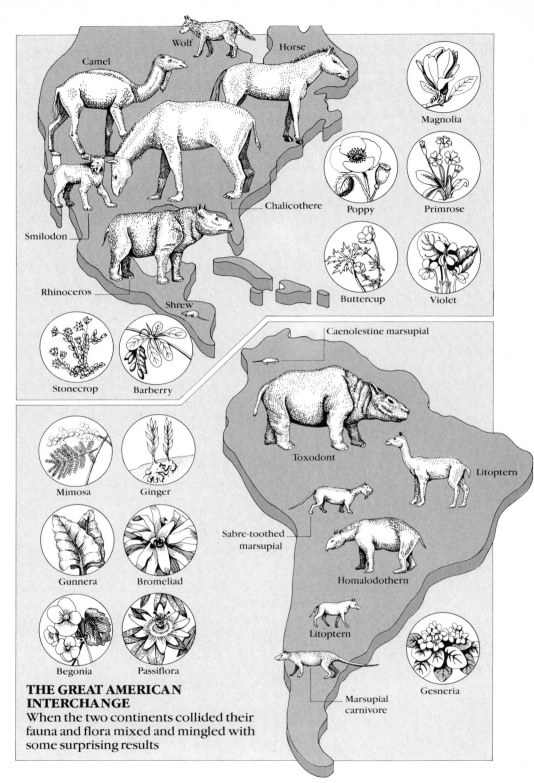

Wolf

Camel

Horse

Magnolia

Poppy

Primrose

Chalicothere

Smilodon

Buttercup

Violet

Rhinoceros

Shrew

Stonecrop

Barberry

Caenolestine marsupial

Mimosa

Ginger

Toxodont

Litoptern

Sabre-toothed
marsupial

Gunnera

Bromeliad

Homalodothern

Begonia

Passiflora

Litoptern

Marsupial
carnivore

Gesneria

**THE GREAT AMERICAN
INTERCHANGE**
When the two continents collided their
fauna and flora mixed and mingled with
some surprising results

The Imperial Mammoth, height to the shoulder as much as 4 metres and weighing up to 8 tons; American Mastodon, height 3 metres; Long-horned Bison with a maximum horn span of 213 cms., over three times that of its modern counterpart; Antique Bison with an average horn span of 88 cms.; Camels, 20% larger than our modern Dromedary; a Giant Ground Sloth about the size of an Ox; a Giant Tapir; the Western Horse and Dwarf Pronghorn Antelope.

All the above were grazers and/or browsers.

Omnivores include the Giant Short-faced Bear, which was larger than the biggest modern Grizzlies from Kodiak Island. With so much heavyweight flesh on the hoof and paw, it is little wonder that large carnivores abounded and what more terrible than a Giant Lion, remains of at least 76 of which are the pride of Rancho La Brea, and a Sabre-toothed Cat as large as a modern lion, more than a thousand of which have been found to date at the same site. To top the lot in number if not in size, the Dire Wolf hunted and scavenged around the tar pits; the same size as the modern Grey Wolf, it was altogether of heavier build. 1,464 individuals of this species have been identified from the excavations.

All these and many more lived in California no more than 12,000 years ago, but a list of names and statistics of size cannot convey the magnificence of these animals, all of which are unfortunately extinct. However, thanks to the vision, generosity and sheer hard work of one George C. Page, you can go and see them in all their glory on the lot bounded by Curzon, Sixth and Ogden Streets and Wiltshire Boulevard. It is one of the gems of the museums and interpretation centres of the world and is worth a lifetime of visits. Highlights for me include the skeleton of a Sabre-tooth Tiger which mysteriously becomes the real thing as you walk past its lair; a wall displaying a cross-section of the skulls of all the Dire Wolves found on site, as ferocious a pack of canine teeth as are on show anywhere in the world; magnificent dioramas depicting the most important events in the pre-history of America; a family of Sabre-toothed Cats; everywhere you turn, a new experience of a vibrant, living past.

However, most impressive of all and central to the whole complex is a laboratory where paleontologists, zoologists, botanists, conservators, designers, artists and many more can be seen at work. This is the real nerve centre of a living, working, museum where new detail is added each day to this unique record of past life. Here new collections from the most recent excavations are cleaned and catalogued and new techniques are used to allow more detailed interpretation of the old. Here you can have the unique experience of peering in through a glass panel at this hub of activity and see scientific endeavour hard at work.

Why did all these animals, and many more, disappear between 10,000 and 6,000 years ago? Was it the final winnowing down by competition after the admixture of the two long separated faunas? Was it the massive changes of climate brought about by the final stages of the last ice age which allowed the re-development of forest both across southern Canada and across central and northern South America? The fact that it was the big and not the little animals that became extinct and that no simultaneous extinctions were recorded in the fauna of the sea seems to indicate that climate may have been an operative factor. Was it some unknown environmental catastrophe which wiped out only the megafauna of the world, leaving the rest intact? Or was it man? To date Rancho La Brea has provided only one firm piece of evidence concerning the presence of man, or rather woman: the remains of a female 25 to 30 years of age who lived some 9,000 years ago and who died suffering from a disease of the sinus bone.

There is now good evidence that man was roaming the landscapes of South America some 20,000 years ago, was in Mexico two thousand years before that and in the Yukon two thousand years before the Mexican date. That he actively hunted and killed big game was first authenticated from a site in New Mexico where in 1926 Dr Jessie Figgins described the bone of a now extinct species of Bison with embedded in it a well-worked flint tool, a Clovis point.

Similar finds across the length and breadth of America both dated and undated leave little doubt that hunting man co-existed with the whole range of the now extinct megafauna. Likewise, finds like that described in intricate detail by Joe Ben Wheat in which a herd of Bison were stampeded into an arroyo, killing 46 adult and 27 immature males, 63 adult and 38 immature cows and 16 calves in a matter of minutes, show that he was well versed in the technology of the hunt. So perfect is the record, the excavation and the interpretation, that we know the kill took place in late May or early June sometime about 8,500 years ago.

This was no isolated incident. Many other such massacres have been found though none so well excavated and documented. Such massive, and we would now say wanton, slaughter, for all the meat could not have been consumed before it began to rot, must have played some role in the extinction of certain species. The Mammoths and Mastodonts had gone probably by 10,000 years before the present, their place taken at least in part by the Antique Bison. Better worked, but smaller Folsom points were the main armour of the Bison hunters. A thousand years later the Antique Bison was gone, its place taken by the smaller *Bison occidentalis*. This species was in turn stampeded to its death by hunters using more advanced stone points which have been called the Cody Complex, and dated 6500 B.C.

None of this is proof that man caused the extinction of any single species, but it does indicate that he may have had a hand with a developing technology of flint projectile points in it. The term projectile point rather than arrow-head is used because none of the hunters had bows and arrows: their weapons of the hunt were simple spears.

Whatever the exact cause, and it was probably a mixture of a number, over a period of some 4,000 years a diverse megafauna was gradually whittled down until when the white man made his first appearance on the Great Plains, the only large grazers left were the modern American Bison and the American Pronghorn, which is unique in that of all the more than 100 similar forms which constitute the family *Bovidae*, it is the only species which sheds its horn sheath each year. For this reason it is put in a separate mammal family all on its own; the *Antilocapridae*.

A vast herd of bison crossing the prairies in the 1860s.

It is estimated that there were probably no less than 60 million Bison feeding on the bounty of the prairies, and the woodlands and semi-desert margins at their flanks. This provided meat on the hoof and a diversity of hunting sufficient to support some 30 major tribes of Indians. The re-introduction of the horse by Spanish immigrants and the proliferation of metal tools and eventually the Sharpe rifle, so increased the hunting potential of both the resident and latterly the immigrant, population that by 1890, the number of Bison had been reduced to a mere 500. Fortunately they were not hunted to extinction. It was, however, touch and go, though since stringent protection measures were enforced in the early 1900s, the total number has increased to more than 25,000. The evolutionary niche of the other 99.96% has been occupied by modern breeds of cattle and all the modern razzmatazz of farming.

This is the dramatic background to the evolution of the diverse grasslands of the prairies. Those which are left in a near natural state owe their diversity and structure as much to this changing panoply of grazers as to the rich soils which support and are supported by them.

But what of the plants themselves? How did they fare when continents collided and the Great American Interchange began?

Unfortunately the records as regards plants are not nearly as detailed. Bones, skulls and teeth not only fossilise well; they are large and diverse in structure and so have many features which make possible their identification. Their discovery and elucidation is also accompanied by a certain kudos for the recorder in question. Leaves, flowers, wood and pollen grains lack all these attributes, at least in part, so the record is far less complete.

We do however know that South America in splitting away from Gondwanaland rafted her own isolated populations of plants west by north eventually to mix and mingle with those from Laurasia. We also know that the plants were more adept at island-hopping dispersal than the animals and so a mixing of the floras took place during some 130 million years before the land bridge was in position. John Raven and Dan Axelrod list 51 families of flowering plants which made the journey north from South America. These include the Mimosa section of the Pea Flower family, and the primitive *Winteraceae*. They enriched a flora made up of members of some 57 families derived from, or at least from contact with, Laurasian stock.

Movement in the opposite direction is less easy to detail, but it would appear that ten families made the southward crossing very early on, while members of 30 more families went the South American way in more recent times.

The difference in magnitude between the south to north and north to south interchange is readily explicable. North America went into partial isolation with a flora that had been enriched by what was essentially a cold- to cool-temperate flora which had made the stepping-stone crossing over Scandinavia and Greenland at high latitude. So, it is unlikely that many truly tropical plant families were well developed or even present. In contrast, South America, though drifting further, had maintained its tropical station, and hence the flora which poured north into the more or less open territory of meso-America.

The cool temperate flora of North America was faced with a broad swathe of tropical climate, whether covered by forest or prairie, which must have formed a barrier to its migration and/or successful landfall. Only when the Andes had risen up into cooler altitudes was the way open for successful migration from the north. For example, Alder appeared in South America only 500,000 years ago and Oak 350,000 years later at about the same time that mountain plants like *Drimys* and *Gunnera* migrated north. What is more, the vast bulk of the 30 families which have migrated south in the last 10 million years are montane (cool temperate) plants. These include members of the Heather, Saxifrage, Buttercup, Primrose, Violet and Rose families.

Oh to be a botanist reincarnated in a few generations' time when much more work has been completed. What will be the story the plants and animals of the prairies have to tell in the light of all that knowledge?

Next morning brought a fine day and an enormous breakfast. It was 7.30 a.m. but the diner was full of cattle people and the talk was of range management and rodeos. It was interesting to listen to their expertise and their concern for the future of their grasslands. Conversations ranged across many topics. One was the dubious success of the Beefalo, a cross between some pretty fancy cows and the Buffalo. The trouble with Buffalo is that it is all shoulder and no hindquarters, an ill-designed animal when it comes to the butcher's block. Cross breeding had overcome that, but fertility was low and there was still a long way to go. Another was a growing concern for the future of the ranges themselves. Vegetarians were pointing out the fact that beef cattle were a wasteful way to use the prairie, for at every step in the food chain, there is an immense loss of energy. A growing body of opinion suggested the removal of the cattle in favour of corn, with beef fed artificially. Much more productive and much more profit. I pointed out that the animals make use of the bulk of the grass crop and require no massive machinery to plough and harvest, nor chemicals and fertilisers. Also that the grain itself in actual fact was only a small part of the total crop; the rest had to be taken into consideration when balancing the prairie budget. I realised that I was preaching to the converted, but felt a little less a dude.

Looking past the menu I read the notice on the wall which listed each and every hunting season. A diversity of game was now managed as part of the overall prairie system. All this would go with the cattle, and the great diversity of the prairies would become a monster monoculture demanding more and more attention and energy to keep it in its prime. Those soft, living soils would be used not as part of a living system, but as a dead substratum on which to grow corn, requiring the increased application of man-made chemicals and use of energy.

That day I walked the tall-grass prairie once more with fuller understanding, and in one landscape unit, saw it all. The foreground lot was already open to the plough, ready to bear a crop for feed-lot beef. In the background, and stretching as far as the eye could see, a diversity of grass was rippled by the gentle wind into a sea of waves. A cowboy sat lazily astride his horse which picked its own way among the tussocks, sending Prairie Chickens squawking on their way. The skyline was marred only by a nodding donkey engine, its pump and storage tank drawing on reserves of energy which were laid down long before there were any flower-bearing plants or warm-blooded animals to interact with them and form these productive grasslands of great stability and beauty.

The immense area now covered with grass had once been covered with water, an ancient sea within the productive waters of which eons of photosynthesis by microscopic free-floating plants kept a complex community of animal plankton well supplied. Their excreta and their remains falling to the sea floor were gradually changed into thick oil, petrochemicals which are now being pumped up to fuel the aspirations of man in the twentieth century.

That afternoon I went to the Osage County Historical Society Museum housed in Old Santa Fe Railroad Depot at Pawhuska station. It is a fascinating collection of bric-a-brac from yesteryear. Old donkey engines jostle for their place amongst poss-sticks, flat irons, family albums, Bell telephones and the useful like. There are rows of sepia-tint pictures of children, grown-ups, grandparents, civic dignitaries, cowboys and the proud members of the Osage nation. The two ladies on duty greeted me with exuberance and answered all my many questions. It wasn't the British Museum or the Smithsonian Institution, but a real living archive of a real live place, the history of real live people who now have the non-renewable minerals and the renewable living wealth of the Tall-grass Prairie in their care: a vast treasure-house with its future firmly rooted in the past.

CHAPTER FIVE

Road from Kamchatka

K AMCHATKA is a part of Russia, as once was Alaska. The latter is now part of the USA, in fact the largest state, and yet it does not figure in the agricultural statistics. The reason is that at those high latitudes the climate is too harsh to support the growth of crops. Its broad landscapes are mainly covered with Entisols, Inceptisols and Histosols. The same is true of the Aleutian Islands but within the limitations imposed by soil type each is as fertile as it can be, the diversity of its flora strictly limited by size. These northern outposts of the USA have much to tell us, but most important is the evidence they contain which relates to the arrival of man in America. For the first Americans walked across from what is now Siberia during the harshness of an ice age, into the boundless lands of the New World.

Attu Island holds the unique distinction of being both the most westerly and the most easterly point of the United States of America. In the knowledge that it is the last (or first) island in the Aleutian chain which stretches, stepping-stone wise, from America to Russia, you can probably work out this ingenious paradox to your own satisfaction.

As the Aleutians go it is a medium-sized island of some 90,570 hectares, its landscapes moderately diverse, rising to a maximum height of 868 metres above sea level. The main problem in assessing its diversity and beauty is that you can rarely see very much of it, because like all of the islands in the chain, it is more often than not shrouded in mist and cloud. No, shrouded is the wrong word, for it suggests something static hanging limply over the landscape. Nothing could be further from the truth because thereabouts the mist and cloud is often moving horizontally, driven onshore by gale-force winds. A sunny day and a distant view on the Aleutians is almost as rare as a tree, for the combination of water-vapour screening the sun and the constant exposure to salt-laden winds prevents the growth of large trees, despite the fact that all the islands lie well to the south of the northern limit of trees on both adjacent continents.

The islands themselves may be devoid of forest cover and hence of forest life but the shallow seas which surround them abound with both cover and life. The cool temperate waters of the North Pacific and the Bering Sea are ideal for the growth of big, and I mean big, brown seaweeds. Take, for example, the Bull Kelp, which, though an annual, can grow to the massive length of 40 metres in one season. Each plant consists of a long thin stipe or stalk held up by a single hollow float. From the top of the float, branched blades hang pendulously down forming the canopy of an underwater forest below which the stipes, each no more than 5 cms. in diameter, snake sinuously down into the depths, at least during periods of slack water. By contrast when tide or current is running they stand angled in the flow, taut as a bow string, tugging at their anchors fixed to the rocks below.

It is an amazing experience to fly bird-like through the sun-marbled canopy, steadied against the roll of the waves only by their buoyant presence, then to dive down into their sheltered world which is only one breath-held dive away. The Germans have a word which at least captures the colour magic of that experience; it is *geldstoffe* (gold stuff), and refers to a complex of pigments which are in the main produced by tiny microscopic seaweeds floating in the water. To a terrestrial being used to moving only in two dimensions amongst the woody trunks of trees, everything about the Kelp plant appears too small, too puny to carry out its func-tion. Yet once below the waves there is no need for tissues for support for the water plays that vital structural role. Likewise no roots are needed to seek out supplies of water and mineral nutrients, for the whole forest is bathed in its own nutritive medium. As far as anchorage is concerned its annual habit protects it from the worst storms of winter and through the earlier part of each year it is growing well below the surface and is thus protected from the strongest effects of swell. A mass of branches the tip of each of which adheres to bare rock is all that is needed to keep the largest Bull Kelp in place. Many is the time that, lungs bursting, I have held and tested their effectiveness before releasing my grip and gliding upwards to the surface, letting the stipe slide through my fingers as I returned to my own environment, exposed to the wind and weather.

If you want the full experience of that underwater world then it is necessary to don an aqua-lung and a good diving suit, wet or dry, and really get immersed in the subject, but do take care, for once you are hooked you are hooked for life.

Sitting amongst the holdfasts of the Kelp, you are surrounded by the ever-mov-ing presence of this sinuous forest, whose shelter forms the habitat for a multitude of other animals and plants. The rock between the Bull Kelp holdfasts is not bare but covered with a felt of other lesser seaweeds, red, green, brown and others of hue less easy to define. These are actively browsed and grazed upon by a whole host of creatures—fish, shellfish, worms, shrimps, crabs, sea urchins and many more—some of which retreat into the safety of the holdfasts when danger approaches in the form of carnivores.

I have sat safe in the protective presence of the Kelp in many different parts of the world and watched a multitude of forest life going about its daily tasks. Kelp forests are one of the few places in the world in which you can see almost the full cross section of the animal kingdom at work, both those with and those without backbones, part of the same society. The list is long and includes both mammals and birds, many now rarities on the world list. Fortunately, the vast majority of the Aleutian Islands were created a National Wildlife Refuge by executive order of President William Howard Taft, way back in 1913.

The most readily visible feature of Aleutian wildlife is its birds, for hundreds of thousands, if not millions, still nest on the islands despite the fact that man has in the past introduced both Fox and Norway Rat into their breeding grounds.

Many of the birds either swim on or dive into the shallow productive waters which harbour the great forest of Kelp and when diving beneath the canopy it is not unusual to have one or two crashing in for a quick fish tea. I find it difficult enough to identify birds when they are flying overhead, but when they are swimming underwater or diving down amongst the Kelp, it is almost impossible, and around the Aleutians there are plenty to choose from.

Amongst the diverse divers are six species of Auklet, three of Cormorants, two Puffins, one Petrel, Fulmars, Blacklegged Kittiwakes, Glaucous-winged Gulls, Guillemots, Murres, Murrelets and Common Arctic and Red-throated Loons. In winter great numbers of waterfowl, mostly Oldsquaws, Harlequin Duck and King Eider, together with half the world's population of Emperor Geese, are found in Aleutian waters. During the summer, cool and windy as it is, Mallards, Greater Scaups, Common Teal, Pintails, Common Eiders and Mergansers nest on the islands, the majority of which once formed the nesting sites for the Aleutian Canada Goose; now the latter breed only on the tiny Buldir Island for this alone was too small when it came to introducing foxes. Today, under wise management, the population is building up and islands from which the foxes have now been cleared will soon be re-stocked. Of shore birds you can take your pick: Black Oyster-catchers, Northern Phalaropes, Rock Sandpipers and raptors like Bald Eagle, Peregrine Falcon, Gyrfalcon and Stellar's Sea Eagle.

Marine mammals abound and most are not uncommon visitors to the deeper parts of the Kelp forest. These include: Northern Sea Lion; Walrus; Narwhal; Grey Whale; and one animal which makes me wish reincarnation in animal form possible, the Sea Otter. It is said to be the animal with the most valued fur coat in the world and for this reason was hunted almost to extinction along this coast. Now under strict protection it has made a remarkable recovery and today more than 20,000 of these gorgeous animals live off the bounty of the environs of the Kelp-fringed coast. They are the real masters of the Kelpland and I have always hoped that one day I will have the pleasure of meeting one at play underwater. Apart from their expensive coat, they have another trait almost unique amongst mammals for like Chimpanzees and Man, they too use tools. It is no more than a large rock, lifted from the bottom and carried to the surface, where they lie on their backs, stone upon tum, and use it as an anvil to smash through the mother of pearl that covers the succulent Abalone, a favourite in their diet.

Despite my predisposition to be a Sea Otter, I must agree that the Aleutians' most famous marine mammal is, or rather was, much larger. Stellar's Sea Cow by name, it was a gentle sea monster weighing perhaps more than two tons when fully grown, the largest by far of all the Sirenia, a group which include the Dugongs and the Manatees, all of which are herbivores and make short work of plants, whether they live floating in or on the water.

Stellar's Sea Cow was the only one which lived (at least within historic times) in temperate waters and hence fed within the bounty of the Kelp, feasting on both seaweeds and turtle grasses alike. Discovered in 1741 on two of the Commodore Islands (now Kommodorski, for they are part of Russia) the population amounted to some 2,000 animals before they were cruelly slaughtered to extinction in a mere 24 years. Georg Wilhelm Stellar was the physician and naturalist on the Bering Sea Expedition which made the discovery and his is the only accurate description we have of the animal and its behaviour.

'Adult sea cows formed rings around young as they fed and "always" kept them in the middle of the herd. Sea cows have an unusual love for each other. Usually entire families keep together, the male with the female, one grown offspring, and a little tender one. When one of them was harpooned, the others tried to save it. Several of them formed a circle about their wounded comrade and attempted to keep it away from the shore, while others tried to capsize the boat which held its tormentors. Still other sea cows lay on their sides and attempted to strike the harpoon and knock it out of the wounded sea cow's body, an attempt that succeeded many times. We also looked in amazement as a male returned to his dead female on the beach two days in a row as if enquiring about her . . .' 'Thirty men were needed to pull a harpooned sea cow on land where it was beaten and butchered while still alive . . .' 'If the animal's mate or young followed they were also slaughtered in this same gruesome way.'

They were killed for their immensely thick hide which was used to cover boats and to sole boots, for their flesh, which was of the finest taste and texture, and for their insulating blubber which when rendered down tasted better than sweet almond oil and helped in the alleviation of scurvy. The last of Stellar's Sea Cows was slaughtered in 1768. In any terms, commercial, scientific or aesthetic, a unique resource was thus wiped out, gone forever, an act of total vandalism. But that was way back in 1768. Surely it could not happen today? If only that were true.

One can only guess what the Aleutians were like before man came upon the scene; a wonderland of nature, teeming with wild life of every sort, a chain of island paradises linking two continents. Now thanks to the vision of President Taft they are safe and perhaps one day will be restored to their full former glory. Unfortunately one thing will be missing, Stellar's Sea Cow, those gentle giants that fed on a variety of Kelp, all of which gained its support from the buoyant salty sea.

Well, all save one, for there is one seaweed which can stand erect when the lowest tides leave it high and drying. Like the Bull Kelp, the Sea Palm is an annual, growing on the lower tidal and sub-tidal zones, especially on shores that are exposed to massive wave action. Each Sea Palm consists of a tuft of fronds, set at the apex of a thick tubular stipe, which, being composed of young rapidly growing tissue, has the texture of tough resilient rubber. It is this rubber pipe construction that makes it unique among all large seaweeds, allowing it to stand erect on its own and weather the storm and anything the wild windy Aleutian weather throws its way. By the end of the season it has, like the Bull Kelp, only completed half of its annual cycle of life for it overwinters as a microscopic filament growing out of harm's way in cracks within the rocks where the worst storms or even grinding pack ice cannot do it any harm.

Standing amongst the Kelp on Attu it is possible to gaze through the almost eternal mists towards the Kommodorskis which, though only some 60 sea miles distant, are at least another world away for they are both across the International Date Line and are part of the USSR. There too in the dim distance of another culture and another day is the peninsula called Kamchatka, a name which is going to follow us clear across the Aleutian chain.

However, before we start on our journey, our road from Kamchatka, is it too much to hope that this new concept of conservation, this sense of caring for the environments of the world and wildlife which they support and which so enriches our lives will unite the purpose of these, the two strongest nations now on earth, in one new purpose? I can think of no better place than the coast of Attu Island on which to stand amongst the grim graffiti of the last world war and the diversity of Kelps which fed the Great Sea Cow, and hope.

The land flora of Attu, consists of 247 species of vascular plant, that is plants which unlike the seaweeds are divided into stem, roots and leaves and having a woody vascular system can stand upright in the air and keep their leaves exposed to the sun and well supplied with all the water they require.

Among their number no less than 208 also grow over on that misty peninsula and four bear its name although in latinised form. They are *Gallium* and *Cirsium kamtschaticum* and *Fritillaria* and *Rhododendron camchatkensis*. Not to be outdone, the flora of this one island also includes 186 species, including some of the above, which also grow over on the Alaskan side. So it would seem feasible to conclude that this island, apart from forming one in a line of stepping-stones between continents, has also been colonised by plants migrating from each of the adjacent continents.

The unique potential of these 70 stepping-stone islands for the investigation of the basic principles of biogeography was realised by members of the Ecology Department of the University of Tennessee and so they set to work in earnest. As part of their programme of graduate studies, Raymond A. McCord started work in 1976 on a study leading to the degree of Ph.D., its title: 'The Biogeography of the Vascular Flora on Islands in the Bering Sea Region'. My thanks are due to Ray and to his mentors Dr Ed Clebsch and Dr Cliff Amundsen for allowing me to delve freely into his dissertation and quote from the detailed results.

The flora of some 29 of the islands were already well-known thanks to the work of the father of circumboreal botany, Professor Eric Hulten. This work was augmented from more recent record and new on-site surveys. Likewise, the flora of six source areas on the adjacent mainlands were studied, the results being stored on computer. Information on the size, shape and landscape diversity of each island determined from large-scale maps was added to the data bank.

The university computer was then set buzzing to answer a number of questions, amongst them, which of the geographical variables show the greatest relationship to floristic diversity of the islands? The answer came in print-out: area; complexity of island shape; distance from source area; landscape diversity: in that order.

One can therefore surmise that sometime in the past the islands were devoid of all vegetation and have gradually been colonised until they have reached the *status quo*.

The first question must therefore be, how do land plants move across a water barrier? Well, those with tiny spores, small, winged or plumed seeds, may be carried on the wind and we have already said the Aleutians are in a very windy place. The seeds and fruits of many water-dwelling plants are adapted to float and like any detached piece of a land plant which can tolerate some degree of immersion in salt water, they can be carried from island to island on ocean currents. Plants whose seeds are borne in juicy capsules or fleshy fruits can be transported inside birds to arrive complete with a wrap-around supply of organic fertiliser, ready potted, as it were. Likewise, any part of a plant which becomes attached to a passing migrant will just be carried along.

The size and shape of any island and the way it is orientated across the wind, current or migration routes, must therefore be of great importance when it comes to intercepting the migrant flora, a fact with which the computer was in full agreement.

All successful land plants invest large amounts of energy in producing a surfeit of such movable propagules and they did it without the help of computer prediction. Once such propagules, each carrying its own genetic message safe inside a survival pack, have reached their new island home, they then enter on their most critical phase of life, ecesis, that is successful germination and establishment. The Bible story of the sower and his seeds sets out in clear detail the immense problems. In natural systems, however, there is one saving grace, for what is stony ground for one species may well be good ground for another, hence the importance of landscape variety, a factor again identified by the computer.

Once arrived and ecesed (what a super word for Scrabble) the whole process starts all over again as the plant in question begins to take over ground on its new island home. From this point on things become much easier as the spread, at least of the perennial plants, comes to rely less and less on seeds and fruits and more and more on detached rhizomes, roots and even budded twigs, offsets, runners and suckers, and all the paraphernalia of vegetative reproduction.

Everything in these once bare Gardens of Eden slowly becomes lovely and continues so to do, that is until one habitat becomes full of plants. Then competition for space and resources really rears its ugly head. One may not immediately think that a plant with its delicate flowers waving gently in the breeze is bursting with aggression, ready to fight for survival. However, any gardener who knows his weeds and takes pride in both his onions and petunias, will understand exactly what I mean. Under natural conditions, the first plants to colonise any open habitat will be those species which are able to nip in quick and make use of the wide-open spaces on offer. However, the eventual winners in the struggle for real estate will be those which are tolerant of competition, in that evolution has fitted them to play a role within a close-knit society of plants, a society which has itself been fitted by community evolution to thrive in that particular habitat, and much more important to make best long-term use of the minerals, nutrients and the solar energy available in that environment.

Thus it is that as the whole situation begins to stabilise and the island, or at least sections of it, become totally vegetated, more plants will go to the compost heap of local extinction and the number of plants present in the flora of the island will gradually stabilise. Whether it ever becomes static is a matter of pure conjecture. Catastrophes like hurricane-force winds, tidal waves and volcanic eruptions, all of which are not unknown on the Aleutians, can wipe out years of vegetative endeavour and create new habitats, new opportunities for new pioneers. Similarly, man can, and does, produce immense changes by introducing either on purpose or by accident new plant and animal species to such a quasi-stable equilibrium.

However, if there is any broad truth in the above scenario, then the area of an island and the number of species of plant growing on it should share a simple relationship and that is exactly what the computer showed for the Aleutians.

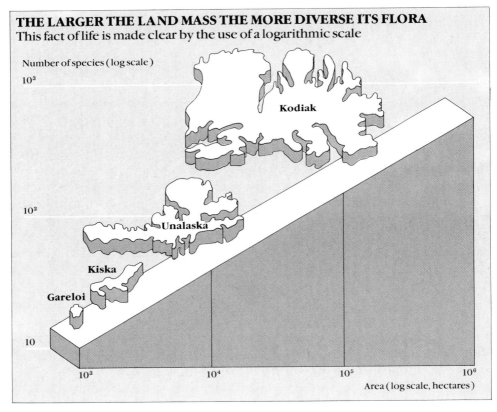

THE LARGER THE LAND MASS THE MORE DIVERSE ITS FLORA
This fact of life is made clear by the use of a logarithmic scale

Number of species (log scale)

Area (log scale, hectares)

Start with the smallest island used in McCord's study—its name Peninsula, its area 156 hectares, its flora six species—pass on to Attu—its 9,057 hectares contain 247 species—and finally the largest, Kodiak, whose 897,400 hectares boast a total of 423 species. What is more, using a log plot (not a mathematical wangle but a way of fitting a range of results into the compass of a simple diagram) all the others really toe the line.

Sit back, look at the diagram and think about it: of course, the larger the island the more plants can grow; but think of the conclusion which is there staring you in the face—the fact *that the potential of* any *land mass to support life is limited, at least in the long term.*

Once you realise that, every landscape takes on a new meaning. You are but a visitor, and if you don't behave yourself an intruder, into a society of plants and animals which have over time come to exploit the long-term potential of an environment to the full. What is more, any changes which are made without due care and attention to these *facts of life* can cause immense damage to the long-term potential and stability of the system. The whole Aleutian experience screams out the same warning message again and again.

What would these islands be like without modern man? The sprawl of concrete, metal and wood left by World War II, which scars the face of many of the islands, is unimportant except to remind us that the root cause of competition and of wars is a battle for limited resources—and ideologies are perhaps the most limiting resource of human nature. These scars of man's adolescence (for our species is a mere four million years of age) will disappear as the natural processes of weathering, soil formation and succession continue. The real shame is that natural barriers have been bridged and feral plants and animals have been introduced which spoil the natural balance. Foxes may have been controlled on some of the islands but what about the rats? Stellar's Sea Cow is no more, as we have seen, and many of the populations of other marine mammals have been culled almost to extinction.

The whole world must mourn their passing but rejoice in the fact that the US Wildlife Service now has the will to restore these islands to much of their former glory, so that here at least the process of natural evolution may be re-established. We must also be aware that the affluence which allowed Ray McCord to obtain his Ph.D.; me to write this book; and you to read it and to sample all the other trappings of twentieth-century life, is also dependent upon the potential of one piece of real estate, called Earth, and more and more, as we shall see, on the potential of America, all 9.2 million square kilometres of it.

On the shores of that great continent at a place called Cape Kreusenstern on Kotzbue Sound in Alaska, there is one of the most ordered of all coastal phenomena: a series of no less than 114 gravel beaches, each of which, starting at the sea, is older than the last. The stretch of coast is aggrading; that is, onshore winds and currents are depositing material so that it is slowly but surely building outwards, reclaiming the sea.

As you walk inland there you are riding a humpback time machine which takes you back through much of the recent history of this part of the New World, for your way, if not paved, is littered with artefacts of human habitation. All are tools of the hunting/fishing trade: the youngest, which include plastic floats and polypropylene, are found closest to the sea, the oldest, dating back 4,200 years are found on ridge 114.

There, this unique roller-coaster history book comes to an abrupt end decked with Millfoil, Arctic Willow, Cow Berry, Jacob's Ladder, and presents its reader with a most ingenious paradox: in order to find evidence of any earlier coastal encampment, it would be necessary to retrace your steps, don diving gear and search beneath the waves. The fact is that it was only 6,000 years ago that the ice sheets of the last glaciation had completed their main phase of melt, returning their land-locked waters to the sea and providing a long-term base for human activity and for the build up of the contemporary coast line.

It took the genius of J. Louis Giddings, archaeologist, to see the potential importance of this site which recorded in great detail more than 5,000 years of Eskimo tradition, a seasonal existence based on fishing the sea and hunting further inland.

Important as these findings were, they also provided Giddings with a pointer as to where to look further inland for sound evidence, and that meant that it must be datable, of an earlier phase of human occupation.

To find such a site in the Arctic is no easy matter, for the environment is not conducive to the preservation of an ordered inventory of past events. For a start, cold temperatures, especially those associated with an ice age, determine that the growth of vegetation and hence the rate of soil and of peat formation will be very slow. This means that an artefact dropped by me in 1980 could fall alongside a stone implement which had been cast aside many thousands of years before. To add to the problem, whatever soil or peat does form is open to the elements and especially to frost action which continually churns and turns the soil, destroying any stratification which might develop. This is even more true in those vast areas of northern North America which are underlain by permafrost. The surface layers of this sheet of hidden ice melt each summer to produce an active layer, refreezing again at the onset of winter. The upper melted layer is not only active in nurturing the tundra plants, but also is subject to constant churning and heaving processes, one reason why the tundra cannot support the growth of trees.

So it might be thought that Giddings was on a wild Arctic Goose chase, but no, the coastal site did indeed give him a clue as to where to look inland for historic Eskimo summer camps. Not only did he know his artefacts, but also his geomorphology and many years before while travelling by raft down the Kobuk river, which flows into Kotzbue Sound, he had recognised a possible stratified site, and this he returned to excavate.

The name of the site is Onion Portage. The reasons for its stratification are complex and needn't bother us here. The reasons why it was long occupied by man are however, much easier to understand.

It consists of a knoll from which hunters could keep watch over the surrounding countryside for the herds of Caribou, Musk Oxen and others. It was also situated on the twisting banks of a river which provided water and transport, an abundant supply of migrating fish, and a watering place which attracted game animals. All in all, a perfect spot in which a hunting/fishing community could spend its summer.

Excavation begun by Giddings and continued after his untimely death, has revealed a fascinating picture of man at the gateway to the New World, his presence

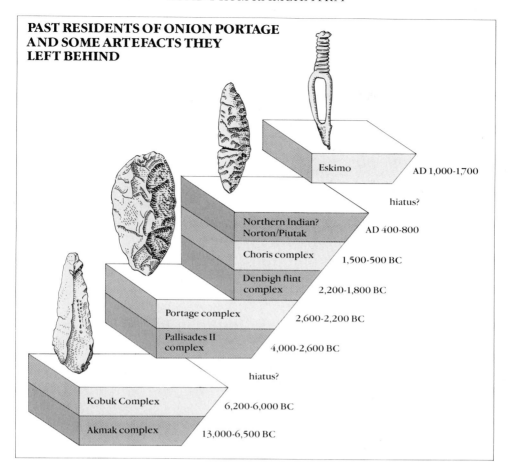

PAST RESIDENTS OF ONION PORTAGE AND SOME ARTEFACTS THEY LEFT BEHIND

Eskimo — AD 1,000-1,700

hiatus?

Northern Indian? Norton/Piutak — AD 400-800

Choris complex — 1,500-500 BC

Denbigh flint complex — 2,200-1,800 BC

Portage complex — 2,600-2,200 BC

Pallisades II complex — 4,000-2,600 BC

hiatus?

Kobuk Complex — 6,200-6,000 BC

Akmak complex — 13,000-6,500 BC

and achievements controlled by the climate and determined by the vegetation of the area. The lowest levels of occupation date from somewhere between 6500 and 13,000 B.C., when the ice age still held sufficient sway to affect both the geography and the climate of the area. Throughout much of the ice age, Alaska, though devoid of ice cover and supporting tundra vegetation, was cut off from the rest of North America by the fusion of two continental ice sheets.

However, to the west stretched the ice-free sub-continent of Beringia, a vast tract of land now submerged beneath the sea, which formed a bridge between the Old World and the New, a bridge that surfaced on a number of occasions, in fact whenever glaciation had depleted the oceans and so had cleared its broad flanks of water and separated the Arctic and Pacific Oceans.

There is little doubt that many plants and animals, including man, made good use of those broad plains, using them both as a place to live and as a route for migration between the continents. Beringia, not the Aleutians, was the road both to and from Kamchatka. That the stone tools found in the lowest strata at Onion Portage show close resemblance to those dating from between 10 and 13,000 B.C.

found around Lake Baikal, in Japan, and in Kamchatka itself, lends substance to this proposition. The tools tells us also that their makers were hunters of a tundra environment, their diet enriched by fish from lakes, rivers and/or the sea.

As the glacial climate warmed up, the sea gradually covered Beringia, shutting off the escape route to whence they came, but at the same time opening up new ice-free land routes to and from the warmer south. The open tundra which had typified Onion Portage for so long, was replaced by Taiga, forests of Spruce, Birch and Larch moving north.

As the environment and the vegetation changed, so too did the people who inhabited the place. In the strata laid down during the warm hypsithermal that followed, the complex of stone tools changed. The local armoury of hunting was not simply enriched by other forms, but changed out of all recognition. Instead of using tools made out of bone, antler and wood, embellished with small stone inserts to give a saw like cutting edge, the main weapons of the new hunt bore projectile points. The new tools include worked stones with notched and shaped bases which must have tipped arrows or spears, flint knives of all shapes and sizes, thin stone scrapers, heavy choppers and notched stone net sinkers. The earliest of these date from 4000 B.C. and the latest, much changed but still of the same broad Northern Archaic Tradition, date from as late as 2300 B.C.

That similar tools were in use at the same time and long before in the southern forests east of the Mississippi is of great interest, and points to a northward migration, not only of the forest but, dare one say, of the cultures that had evolved within its affluence. Much more archaeological evidence is required to substantiate such claims, but the findings at Onion Portage point in that direction.

The story Onion Portage tells does not end there, because strata closer to the surface reveal another phase of occupation which links directly with the layered beaches at Kotzbue Sound.

The artefacts of the 'woodland' people are now replaced by what is recognised as a very widespread Arctic Small Tool Tradition, which centred once again on edge-insert tools. These were similar to those of the earlier phase, though they were undoubtedly made by more skilled craftsmen, for each insert has been delicately flaked into a half-moon shape. The people who made these lived on the open tundra, which had now once again replaced the forest, repairing to the coast each autumn to camp and harvest the riches of the sea.

It is perhaps too simplistic to state blandly that the forest people drew south with the retreating forest, to be replaced by hunters who came back to reclaim their former hunting grounds, but this is what is inferred.

The Onion Portage site now stands silent, just within the northern fringe of the Taiga forest belt. No hunters sit there looking for game and no archaeologists brave the hordes of biting insects to wrest the pre-history of America from the silt sandwich that lies beneath the stunted trees.

This is one of the world's great archaeological sites, ranking alongside Tepe Sarab, Ur of the Chaldees, Stonehenge, Machu Picchu, Pompeii and many more in terms of what it tells us about ourselves. I am sure the archaeologists will return.

The welfare of hunting man is dependent upon the vegetation. So too is that of all the wildlife that feeds on it, or gains shelter from its presence. This interdependence is not just a one-way effect and grazing pressure may well be an important factor in determining the rate of advance or retreat of the forest edge. Recent research at the University of Alaska has shown that these relationships are both closer and more complex than was originally thought. John P. Bryant's work centres on the Snowshoe Hare, whose white winter coat and hair-covered feet provide protection from at least two aspects of a snow-clad landscape. Despite this, its population seems to fluctuate in a somewhat dramatic way over a cycle of approximately ten years with spectacular crashes following a gradual build-up in numbers.

The Snowshoe Hare is both a grazer and a browser and as winter approaches, the accumulating snow cover lifts it further and further up the trees whose succulent twigs, which are full of stored winter food, become the major factor in its energy intake. Work has shown that they are fastidious and there are species they prefer to eat, like Willow, Aspen and Birch, and others, like Alder and Black Spruce, which they avoid. It is of interest that the former are pioneer trees of the forest-tundra border whereas the latter are dominants of the stable Taiga forest.

The secret of the success of the latter appears to be unpalatability, which is not due to low nutrient value but to secondary plant products such as resins which evidently makes them taste nasty even to a hungry hare. In fact the foliage and floral buds of the Green Alder, Alaskan Paper Birch and Balsam Poplar, which are all rich in carbohydrates and proteins, are completely rejected by the hare for they are also rich in resins.

To prove the point, Bryant coated bundles of the most palatable shoots with extracts of the resin and found that he could obtain complete rejection, even by the hungriest of hares, with a concentration of resin less than half that found in nature. It therefore seems a tenable hypothesis that these secondary plant products now act as chemical defence mechanism, protecting the plants from overgrazing.

To simplify a very complex picture, we will stick with the resin of one particular type of tree, the Paper Birch. In those parts of North America where browsing by Snowshoe Hares is concentrated, the population of Paper Birches is characterised by resin glands on the internodes of all immature twigs, and these shoots are rejected by the hares. Add to this the fact that the resiniferous bud scales are also unpalatable and the poor old hare has got a real problem and the Paper Birches are put at great advantage. However, as the twigs mature, the picture changes, for they lose their resin glands and become palatable and indeed in places become the staple diet of the hare, which has to stand on tip toe to reach the resin-free goodies.

If such browsing is too heavy, then the Paper Birch begins to suffer. Its mature shoots nipped off in their prime, each tree responds by producing new adventitious shoots from round its base, young shoots which almost drip with resin and are rejected by the hares. The trees thus live to mature a year later and under extreme cases the Snowshoe population crashes, only to rise again once there is a new abundance of mature tasty shoots, by which time leader shoots may well have got away to produce a standard tree out of reach of even the tallest hare.

The real twist in the tale of the Snowshoe Hare is that like so many herbivores, it is dependent upon a whole gut flora of bacteria which help to digest the tough woody tissue. Certain chemical constituents of the resins we have mentioned have been shown to have an anti-microbial action against those very same bacteria.

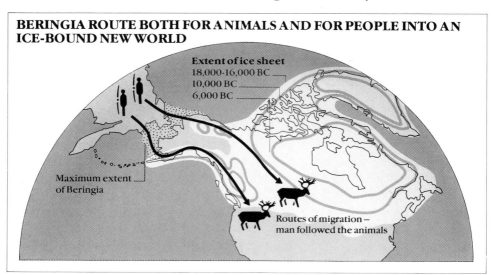

BERINGIA ROUTE BOTH FOR ANIMALS AND FOR PEOPLE INTO AN ICE-BOUND NEW WORLD

Alaska, though a tough, harsh environment, is one of great beauty and diversity, stretching as it does across the Arctic Circle into the light of the midnight sun and west, or is it east, to Attu. It is a land of mountains, glaciers, coastal plains, rivers, estuaries and a rich sea coast, a land of great potential, as the new settlers of Anchorage, Fairbanks, Juneau and a scatter of other vibrant townships know.

It was the gateway to America for man and many animals. The two halves of the great super-continent of Laurasia, which had drifted apart were there re-united, once between 50 and 40,000 B.C., and again between 18 and 8000 years B.C., to produce a broad migration route. The secret of exactly when, and perhaps even where, man made his first crossing into the New World may still be locked in that sandwich of soil and river gravels waiting to be excavated at Onion Portage. In all, a rich pre-history, stretching back towards the dawn of human kind, just part of the natural heritage of the United States of America, and of its neighbours Canada and Mexico.

CHAPTER SIX
West Side Story

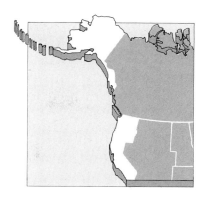

MAN entered the New World across Beringia, the soils of which have long since disappeared beneath the sea. As the ice sheets of the glaciation melted, two routes were opened up to his nomadic talents: one down the centre of the continent through an ever-widening ice-free corridor, the other down the coast, along the west side of range after range of mountains covered with dense forest and capped by melting glaciers. With no more than simple stone tools there was little men could do but hunt the edges of the forest and so they would have been kept on the move. However, certain locations which offered something special, something which eased the burden of constant nomadic wanderings, became focal centres for the development of new cultures. Excavations are slowly but surely revealing the story in detail and are demonstrating that man is subject to the same rules governing opportunity and limitation as the animals and plants on which he depends. Likewise, a study of Indian cultures which existed well into historic times is helping us to understand the problems of a life depending on the sea and the Histosols and Ultisols of the coastal strip.

The folklore of many nations across the earth includes stories of a great flood which either pushed their ancestors from traditional hunting grounds or, in its extreme form, made the wise take to arks and float to pastures new. Far-fetched as certain details of some of these may seem, there is little reason to doubt that they refer to the great inundation of vast areas of land as the melt waters of the last or the penultimate glaciation returned to the sea after their sojourn on land.

I add the word penultimate in the context of this story of the New World because sound evidence shows the presence of hunting man, with Asian connections, in South America as early as 20,000 years before present. This certainly suggests that man crossed Beringia long before the last melt opened up the southward routes, that is either during the build up of the last ice sheet or during the previous one. If this is true, it would push man's discovery of the New World back to at least 40,000 years ago. On the evidence to hand I would plump for the first, but the exciting thing is that one day we shall know. New sites, new excavations and new techniques are providing the answers and will prove some theories right and some wrong. This is the real excitement of America, for anywhere you are you could be standing near some prehistoric site and, because it is prehistoric, there is nothing in the way of written evidence to suggest its presence.

Whatever the exact date of the first crossing, at some time the people of Beringia must have been faced with rising damp as, over a period of some 4,000 years, their subcontinent was divided into peninsulas, islands and eventually disappeared beneath the waves, leaving them stranded either in the Old World from whence they set out or the New World which was wide open for colonisation.

It seems also safe to suggest that those groups which included fishing in their life-style would have stuck to the coast, making good use of everything it had to offer. The fact that the whole richness of this Pacific coastal strip was flanked on its landward side by high mountains (in order proceeding south, the Alaska Range, St. Elias Mountains, Coast Mountains, Cascade Ranges, Coast Ranges,

NORTH WEST COAST FISHING

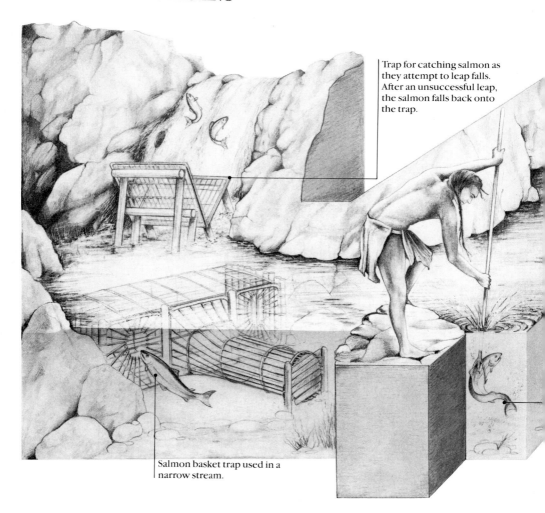

Trap for catching salmon as they attempt to leap falls. After an unsuccessful leap, the salmon falls back onto the trap.

Salmon basket trap used in a narrow stream.

Sierra Nevada, in the main backed by the Rocky Mountains themselves) made it very much more a route for settlement and southward migration than a way

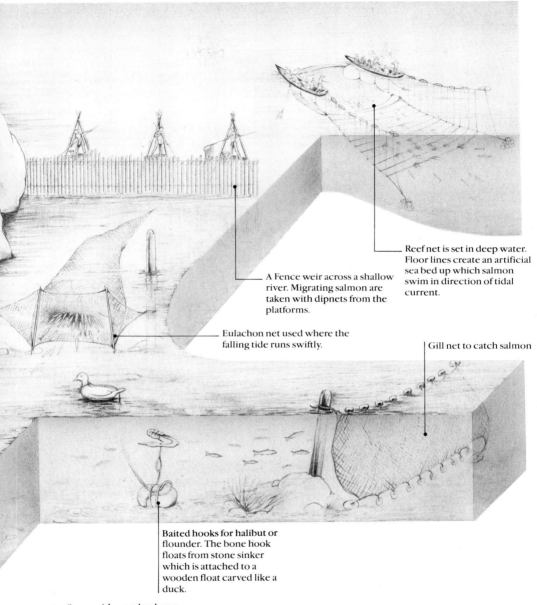

Reef net is set in deep water. Floor lines create an artificial sea bed up which salmon swim in direction of tidal current.

A Fence weir across a shallow river. Migrating salmon are taken with dipnets from the platforms.

Eulachon net used where the falling tide runs swiftly.

Gill net to catch salmon

Baited hooks for halibut or flounder. The bone hook floats from stone sinker which is attached to a wooden float carved like a duck.

Spear with wood or bone barbs used for catching fish or waterfowl.

inland. Having said this, there is little doubt that the hunter fishermen would have travelled inland with migrating fish as they moved up river to spawn in the river headwaters, headwaters that were fed at least in part by melting mountain glaciers which would have themselves posed another barrier to inland migration.

There was another inland route which opened up between the two main melting areas of the continental ice sheet at the same time. This led south-east through what is now the Mackenzie River Valley to the Great Lakes, which were then emerging from beneath the southern limit of the ice. For the time being we will, however, stick to the western route and its story.

Lying across the coast route south in what is now called British Columbia are a wedge-shaped group of some 150 islands called the Queen Charlottes. Their combined land mass of 3,600 square miles is separated from the mainland by the Hecate strait, which is at the most 100 metres (300 feet) deep. There is no reason, therefore, to doubt that at the height of the glaciation they were all joined and formed a peninsula, an extention of the mainland which was covered by a lobe of the continental ice sheet. The exact extent of the ice cover is, however, a matter of conjecture and debate.

Cattle roam the beach on one of the many Queen Charlotte Islands.

Some say that only some 3.5 square miles of land stood proud of ice as rocky nanataks, upon whose cold and ice-blasted flanks nothing could grow. Others argue that on the then steeply shelving Pacific coast (now hidden by the sea) there were extensive areas of open land which formed a refuge for a diversity of plants, from which they spread to the adjacent mainland as the climate improved.

Key in the latter argument is the presence of no less than eleven plants, over 2%, out of a contemporary total native flora of 472 which were until very recently thought to be endemic, which means they grow nowhere else in the world.

They include a grass, *Calamagrostis purpurescens* ssp. *tasuensis*; a lily, *Lloydia serotina* ssp. *flava*; a violet, *Viola biflora* ssp. *carlottae*; a heath, *Cassiope lycopoides* ssp. *cristapilosa*; a willow, *Salix reticulata* ssp. *glabellicarpa*; a lousewort, *Pedicularis pennellii* ssp. *insularis*; an avens, *Geum scolfieldii*; and a buttercup, *Isopyrum savilei*. I apologise for all the Latin names, most of which must seem longer than they need be, but it is their trinomial nature, rather than their binomialness which is important to the argument. The third name, prefaced by ssp., means that the plant is only of subspecific rank, and thus bears great resemblance to species whose distribution is not restricted to the Queen Charlottes. Only the last two, the avens and the buttercup, rank as true endemic species.

It is argued that the process of evolution of a new species takes time (much longer than that which has elapsed since the melting of the ice) and so the ancestors of these plants must have lived on such coastal habitats throughout the glaciation. There is little doubt that the coast of the Queen Charlottes could have provided such habitats. If only all the eleven endemics were of full specific rank; if only we knew more about their distribution pre- and post-glaciation, if only . . . well, we don't; but again, who knows what future research holds. The recent discovery of all these plants growing on a remote area of Vancouver Island not far to the south, does not invalidate the above argument but shows that there is still much work to be done and new things to discover.

Suffice it to say that, as the ice cleared from the land and Hecate Strait became filled with water, the broad diversity of the Queen Charlottes was there for the taking.

That the warmth of the hypsithermal period saw a marked enrichment of the local flora is evidenced by a number of plants which today find their northern limit in America on these islands. It is especially interesting that three of them, a jacob's ladder, a geranium and an anemone, are found only along the coastal bluffs of a limestone island where they thrive in the warm, rapid-draining soils on which few other competitors can grow. It is there, with the sea on one side and mixed rain forest not far inland, that you can really get the idea of vegetation held in dynamic balance by a changing environment.

Today the environment of the Queen Charlottes and of much of the coastal strip is oceanic in the extreme, an alternation of cool summers and mild winters, well washed with an annual rainfall of between 100 and 750 cms, depending upon location. This is an ideal environment, both for the growth of peat and rain forest and, to prove it, that is what covers most of the area: a rich tapestry of vegetation made up of no less than 472 native species which, for an area of 932,400 hectares, falls neatly within the predictions of the McArthur and Wilson hypothesis and the Aleutian experience.

To a peatnick, that is, someone who has been studying peatlands (or, as they are known in North America, Muskegs) all his academic life, the eastern lowlands of the Queen Charlottes are pure wet heaven. Take a walk from Tllel on the east coast to Port Clements on the Masset Inlet and you traverse a wonderland of bogs and mires, a semi-solid, part floating, magic carpet, each undulation in which has its own story to tell.

I can feel the water oozing between my toes, for barefoot is the best way to go, and smell the aromatic sweetness of Sweet Gale upon the evening breeze as

I revel in the memory of my favourite plant, the Bog Moss, *Sphagnum imbricatum*, for it dominates the scene. Before the depredations of man brought atmospheric pollution and regular fire to the mires of Britain it was a common species, in some places covering vast tracts of land with its golden-yellow to orange-brown spongy presence. Today it is a rarity restricted to a few remoter spots. I have followed this plant across the world, hoping to see it playing its true role as the major manufacturer of its own peaty substrate, and it was on the Queen Charlottes that I first fulfilled my wish.

The wet oceanic climate, in which precipitation never lags far behind evaporation, makes it an ideal place for the growth of Sphagnum and hence the development of a thick blanket of peat. Though this usually has its initiation in lakes, pools and ponds, given sufficient rainfall and flat undulating topography, with no sharp breaks in slope, the peat can start to form almost everywhere, coalescing to form a living blanket which supports a very special community of living things. The development of peat is in effect an extreme case of Podsol formation, in which the constant leaching of rainwater not only removes all the available nutrients but also waterlogs the leached soil layers, preventing oxidative decay of plant litter. An organic blanket, peat, thus begins to form, further aggravating the problems of mineral supply from beneath, so that in time the only nutrients available to plant growth are those held in cycle and the meagre supplies brought in by the rain. Furthermore the accumulation of organic acids produced by the plants reamins unneutralised, souring still further the waters and the habitat. The result is that very few species of plants are able to live along with the Bog Mosses.

Such acid peatlands cover vast tracts of the earth, being particularly abundant in the wetter, cooler latitudes of the northern hemisphere. It is indeed possible to walk clear across North America and, come to that, Eurasia, and never be far from peatlands and the plants which given them their character.

I can never quite get over the fact that you can stand in a peat bog on the Queen Charlottes and realise that the vast majority of the genera and many of the species of plants around you are found in similar associations, in similar habitats, in the Yukon, the coasts of Hudson Bay, Newfoundland, Ireland, Scotland, Scandinavia, France, Germany, Poland and the breadth of Siberia, clear across to Kamchatka and many places in between. These are just some of the immensely successful circumboreal flora which have benefited from the drifting of the northern continents into the cooler latitudes with the consequent formation of extensive peat. When it comes to being productive on minimal mineral input, these peat-producing species rank amongst the most 'switched on' plants in the world, and the Bog Mosses are top of the league, with *Sphagnum imbricatum* undoubtedly king of the habitat.

In the extremely wet environment of the Queen Charlottes, wherever the peat blanket covers sloping ground, or where it has itself grown into great domes called 'raised bogs', it often splits and tears, producing pools which may be trench-like and arranged at right angles to the slope, forming patterned surfaces which are best seen from the air. At ground level these pools, which are full of water stained dark brown with humus, bar the walkers' way and are treacherous, but only to those who do not know the ways of peat and the habitats of the plants that grow to make it. *Sphagnum imbricatum* is a tough, chunky knit plant which always provides firm footing, or a firm haul out once you have decided to take a swim in the dark, summer-warmed waters. *Sphagnum cuspidatum*, on the other foot, is of much finer structure and always warns of a soft spot, in all probability the final stages of infill of one of the pools.

The Bog Mosses thus form a living carpet, an organic cover-all, except in places where water movement enriches the habitat with nutrients brought in from mineral ground and allows other types of moss and sedges to dominate the scene.

Without doubt, my favourite spots are the pools, for there you can really get immersed in your subject and, at the same time get most of you away from the hordes of biting insects which patrol the peatlands on windless days. The water, being dark brown, absorbs the heat and the upper layers warm very quickly and stay warm, in marked contrast to lower layers which feel ice cold. Each pool, unless it is supplied with a strong mineral-rich throughflow, is usually rimmed by a floating mat of Sphagnum upon which Sundews grow in such abundance that the sticky leaves of several adjacent plants can trap even a fully-grown Dragonfly. There it will remain, its tissues part rotting, part digested by enzymes secreted from the leaves of this insect-'eating' plant, which thus gains precious nitrogen, the least abundant nutrient in cycle in this acid habitat.

Dangling down from the floating mat, three other insectivorous plants, Bladderworts by name, may ply a similar trade, but this time under water. They catch their prey, which includes minute insect larvae, crustaceans and many more, by means of hydrostatic traps, each one like a miniature plastic bag. The open pools themselves are habitats for the true aquatic plants, the most conspicuous of which is the Yellow Water Lily with large yellow flowers and floating leaves or lily pads. This is the same species as that which grows in Europe, *Nuphar lutea*, but its cylindrical flower stalk and sepals which are tinged with yellow within signify it to be the subspecies *polysepalum*.

As the floating Sphagnum lawn grows out over the pool, slowly filling as it goes, the water lilies get squeezed into smaller and smaller compass, until what remains of each open pool is chock-a-block with lily pads, which are so crowded that they grow packed vertically, unable to float on the surface. Finally, as the Sphagnum mat continues to overgrow the habitat, the condition for the continued growth of the water lilies gets so bad that their leaves become scaled down to bonsai miniatures of their former selves until at last they disappear from view, a mere memory of the open water, their remains safe in the peaty record. With them go many other plants and animals which found their homes in the uncrowded waters of the open pools.

One of the main species of fish which inhabit the larger lakes and ponds of the flat bogland is a species familiar to all children across the northern hemisphere: the Three-spined Stickleback, which when adorned in all its armoured glory has three sharp spines along its back and two short spines protruding from the underside of its pelvic girdle. This physical structure is usual throughout its range, and so this particular fish can be readily identified and caught with a simple net.

In the boglands of the Queen Charlotte Islands things are not so simple. In fact, each lake appears to contain its own physically distinctive variety of stickleback. These range from monsters 20 cms long, dark in colour and bristling with massive spines, down to diminutive specimens which hardly have any spines at all.

Work at the aptly-named Drizzle Lake Reserve near Port Clements is investigating this diversity among the sticklebacks, and their findings are suggesting some very surprising conclusions.

The main correlation between the spininess of the fish and the habitats in which they live appears to be the type of lake, and the predators which are able to live within its confines.

The open lakes, the ones with both inflow and outflow streams, support populations of Cut-throat Trout, Rainbow Trout, Dolly Vardens and Sculpin. The smaller, closed, lakes which are totally surrounded and often undergoing colonisation by Sphagnum, lack populations of these fish but abound with Pond Beetles and the nymphs of large Dragonflies, all of which eat sticklebacks. All the lakes are also visited by a variety of fish-eating birds: Kingfishers, Red-necked Grebes, Horned Grebes, Hooded Mergansers and both Red-throated and Common Loons.

Tom Reimchen, who lives and works at Drizzle Lake and is one of the most colourful characters I met in North America, lent me the following field notes and his collection of sticklebacks from which the drawings were made.

| | | PREDATORS | | |
	NAME OF LAKE	INSECTS	FISH	BIRDS
Small closed lakes	Serendipity	●		●
surrounded by muskeg	Boulton	●		●
with least armoured	Nuphar	●		●
almost spineless sticklebacks	Solstice	●		●
	Kumara	●		●
	Drizzle		●	●
Intermediate lakes	Eden		●	●
	Mayer		●	●
	Smith		●	●
	Wegner		●	●
Larger more open lakes	Poque		●	●
with spiney armoured	Dead Toad		●	●
sticklebacks	Darwin		●	●

The evidence looks almost too good to be true. Predatory fish, however ferocious, would find it hard to swallow a large, fully-spined stickleback. Fish-eating birds would face similar problems. However, insects which hold on to their prey while using their efficient mouth parts to devour its body piecemeal would find the spines a help rather than a hindrance. The bigger and better the spines, the easier to hang on to their meal.

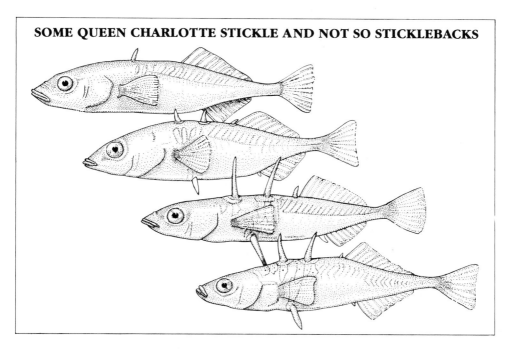

SOME QUEEN CHARLOTTE STICKLE AND NOT SO STICKLEBACKS

Thus the spiny sticklebacks must draw considerable advantage from living in the larger, more open, lakes, while the less spiny forms would benefit from life in the weed-infested waters of the lakes at the top of the list.

If this is indeed true, and there is still a lot of work to be done to prove or disprove it, Tom Reimchen and his co-workers may well be witnessing evolution by natural selection in rapid action—rapid, because we know that 8000 years ago the lakes did not exist, the Queen Charlottes were then only just recovering from the effects of the weight of the ice sheet and whole sections of the islands were still beneath the sea. Sticklebacks can live in salt water, and so it is suggested that as the land rose beyond the reach of the sea, each lake could have come into being with its own population of uniform sticklebacks, from which the contemporary diversity of forms had developed. If this is proved to be true, and the work is going on apace, the Queen Charlotte Islands will rank alongside the Galapagos as heritage showcases for Darwin's theory of evolution.

We shall probably never know whether a new spineless species darting unfettered through the weed-infested waters will go to full fruition. The problem is that in this wet environment continued peat growth eventually destroys the habitat which favours the spineless fish. Predatory birds can move on, but their prey are doomed, like the water lilies, to an ever more crowded existence and eventual extinction. Mind you, fish can be rescued by avian Noahs flying from pool to pool, dinner 'in beak', who can drop their prey, still alive, into a larger pool en route. The trouble is that they are much more likely to drop the ones with spines than the ones without. Such is the way of evolution and of life.

What of man, who found himself marooned along with the sticklebacks on the Queen Charlottes and along the adjacent coast at about the same time? The earliest artefact found dates back to around 10,000 years B.C. That they, too, or at least their direct descendants, witnessed a great flood is recorded in many of the folk tales and depicted on the totem poles which still adorn the coast. My favourite tale concerns the tribe that befriended a whale and climbed into the safety of its mouth until the flood had subsided; others claim simply to have climbed their totem poles and sat tight. Whether these tales refer to the post-glacial rise in sea level or to the effect of tsunamis (tidal waves) brought about by the earthquakes which are not uncommon along this coast, we do not know. We do know, however, from the artefacts they left behind, that like the birds and insects which ate the sticklebacks, they too were fish eaters. What is more, they still relate the following story of how the sea on which they depend came to be filled with fish.

'And Raven he begin to draw it into shore little by little. Finally he pull it onto the beach and he jump inside and he open each door. He open the doors for smelts (fish, small fish) and the smelts comes out of the tank. After that herrings and oolichons, and out of the other sides, king salmon first and humpies and coho, and later on the ones they call the fall fish, dog salmon, and last comes the ones that stop, the halibuts and flounders and cod, and he pushed them all out.

And Raven was satisfied, he released all that fish to go around the world.'

The rich hunting grounds of the sub-continent of Beringia, teeming with Caribou, Musk Ox, Mastodon and Mammoth, were thus slowly but surely replaced by rich fishing grounds. New homes and encampments were located on peninsulas, which became islands; and the quiet coastal waters of sheltered straits and bays provided proving grounds in which to test new skills of aquatic survival.

These new inshore waters were the nursery grounds for these new cultures of man, and served the same role for the fish which came to form their staple diet. 'The ones that stop', Halibut, Flounders and Cod, are the ones that do not migrate; that is why they had to be pushed out of the Raven's legendary box. The others are the migratory fish which run up river each year to spawn. As the river systems of the sub-continent of Beringia disappeared beneath the Bering Sea each species of fish·had to find new rivers in which to ply their anadromous way of life. As the ice melted, the rivers Fraser, Nimpkish, Bella Coola, Skeena, Kitmal, Nass, Kimsquit, Stikine, and many more came into being, each fed by the dwindling glaciers of the coast range. Each were short-run rivers, arising in the coastal mountains and tumbling headlong into the sea, their turbulent waters well oxygenated and thus providing ideal conditions to incubate multimillions of eggs. So the fish came, and still come, each year in the order set out in the legend: Smelts, Herring, Eulachon—and out of the other side of the box the big predators which would have eaten the smaller fish had they been together in the same compartment—Sockeye, Chinook or Spring, Pink or Humpback, Coho or Silver and, in the autumn, Chum or Dog.

In these modern days of hypermarts and freezers it is, perhaps, difficult to understand the importance of such seasonal bounty, a plentiful supply of fresh protein swimming past your home throughout much of the year. Once the skills had evolved to tap the spawning wealth, there was no need to worry, except through the long winter, and with the right amount and the right type of preparation there was no need to worry even then, whatever the winter weather threw your way.

Most important, however, were the first fish to run up river, especially if the winter had been harsher or longer than usual, and most important of all were the Eulachon, also called the Salvation Fish. They came to spawn, countless millions of them, in all rivers in which the tide runs strongly. Starting in the north in late February, which is the hungriest time of the year, each migration lasts about two weeks, ending in the southernmost rivers in April: six to eight weeks of oily plenty, for another name, the Candle Fish, refers to their prodigious oil content—supplied with a wick, a dried fish can be lit and will burn like a candle.

It is not difficult to imagine the scene as the first silver armada streamed into the still frozen Nass river. Fishing was carried on through holes cut in the ice, through which a simple bag net was inserted, its fork-like prongs pushed into the river bed. Sea Lions, Porpoises and Whales followed the Eulachon, gorging on their plenty, while Seagulls and Bald Eagles flew overhead, the former in raucous hordes, the latter in majestic soaring solitude, each taking its share alongside man.

The first run of fish were eaten raw, washing away the taste of dry smoked meat which had pervaded their lives throughout the cold months. Others were dried for the months ahead, but the vast majority were dried prior to being rendered down for their oil, a rich source of energy, vitamins and taste, a food cure-all, condiment and even a cosmetic par excellence, there just when it was needed most!

The twine for making the nets came from the stem of *Urtica dioica*, the common Nettle. Like the Three-spined Stickleback, the plant is circumboreal, as is the technology which developed around its strong fibres. The Nettle is a common plant of enriched soils at the edges of rivers and in forest clearings. A phosphophile by nature, it enjoys the company of man, growing on middens (garbage dumps). It became a useful addition to his way of life and a welcome and useful crop each year. It would be interesting to know how human beings which, thanks to the Bering land bridge, had themselves become another highly successful circumboreal species, learned to brave the stings, to cut, split, dry, peel, pound, shred and spin the fine nettle twine. We shall probably never know. We can, however, surmise that, necessity being the mother of invention, with estuaries teeming with tasty fish and long hard winters to sit and think, they grasped the nettle and want was not their master.

The other sources of twine all came from larger, woodier plants: the bark of young Hooker's Willow, Indian Hemp; and most widespread, and hence most important of all, the inner bark of the Cedars, both Yellow and Red, although the former was preferred because it was both finer and softer in texture. Again, great effort and skill were required to collect, prepare and to get the best out of this inner bark, techniques evolved by trial and error over an immense length of time.

The cedars were also used in another way for catching fish, even before they were hatched. Leafy branches suspended in the right part of the sea, at the right time of the year, attract the spawning Herring which cover the branches with their eggs. The result: an energy-rich confection, a cross between caviar and aspic tinged with cedar aromatics.

Certainly the strangest form of fishing line was provided by the sea itself. The longest rope-like stalks of the Bull Kelp were collected, dried, cured and then either spliced or plaited together to make very strong lines, each as long as 200 fathoms. They were stored dry and became brittle, but immersion in salt water soon put back both their pliability and their strength.

Using a series of such lines tied together with the circumboreal fisherman's knot, the Haida, the people of the Queen Charlottes, tackled the deep waters for giant Halibut, at least until recent times which introduced less exotic but, unfortunately, less bio-degradable lines. In the old days nothing went to waste, for even the hollow floats of the Bull Kelp were kept and used as containers for Oil of Eulachon. The importance of this oil, which the locals called 'grease', cannot be overemphasised for, though their landscapes produced fish in abundance, it produced little in the way of chips, no abundant source of carbohydrate to go with the protein and none which could be stored for use in winter. Grease was thus their main source of energy and hence internal heat.

Imagine what life would be like without flour, rice, potatoes, turnips, carrots and the like—and indeed without sugar. Grease was their year-round answer, since only in summer was it possible to sweeten the diet with Salal Berries and the underground stem of Liquorice Fern.

Despite this lack of plant staples, the plant products of both forest and field were of great importance in their diet, medicine and technology. Nancy J. Turner's fascinating and definitive works on plants used by the coastal peoples lists some 200 which were used for food medicine and other purposes.

Within the limits of this book, all I can suggest is that you consider a wealth of plants 'woven' into the intricacies of a way of life. Then if you can, please go and see it for yourself. Wander through the forests dripping with the promise of rain just fallen, and across the patchwork of the peatlands, look at the giant trees standing guard over lesser plants, flowers, ferns, mosses and liverworts, then look out to sea. Think of all those fish running out of reach until the inventive genius of the descendants of those first men and women who crossed the Bering Bridge and the mother of invention, in time and above the reach of the highest post-glacial tides, put together a rich new way of life, so rich that within it they found both time and motivation to develop a religious way of life and an art form, the like of which is found nowhere else on earth. Both are for me epitomised by the crafting of a traditional kerf bent box. A Cedar was selected in the forest both for girth, at least three metres, and straightness of growth. After seeking permission from the spirit of the tree, two deep holes were bored in the trunk, one near the ground, the other high up, a little higher than the length of wood required. Planks were then split off using a graduated set of wedges made from Yew wood and a mallet made from stone, the tree and its spirit being left to live on in the forest.

Two such planks were all that was required to make a box. The shorter one which formed the bottom was grooved or rabbeted all round to receive the sides. The other, longer plank was then deeply kerfed (that is, grooved in one of a number of very special ways) before being steamed until plastic. Stale human urine was also used to help in the plasticising process, and then the board was bent to form the four sides of the box, the open corner being sewn or pegged together. Set into the rabbets on the bottom board the resultant box would hold water, or indeed anything else, whether utilitarian or ceremonial. The outside of the box was ornamented with carving of the highest order, the designs being the crests of the owner, which verified both rank and spiritual power. Finishing touches were put on with Dogfish skin used as sandpaper, dermal denticles (that's what makes it rough) on the outside.

Boxes could be bent into any shape to fit in the prow or stern of a canoe, which may itself have been produced much in the same manner by the same craftsman. However, perhaps the most important and the strangest use was to hold the most precious possessions of the owner for transportation to the Potlach.

Potlach is a Chinook jargon word meaning 'to give' which has come to be a

Cedar box made by Haida Indians, like this woman photographed in 1900.

cover-all term for a whole feast of separate ceremonies which involve two main principles: firstly, that all events of social or political interest must be publicly witnessed; and secondly, that all those who perform personal or social service must be recompensed. There is a third important part of any potlach: everyone present on such occasions must be fed, which is why I used the phrase 'feast of ceremonies'.

So it was that when time or an event such as a betrothal, marriage, death or some other important happening occurred, acquaintances were invited to the feast at which goods were distributed. The larger and more widespread the number of guests, and the more lavish the presents and the feasting, the greater was the prestige of the giver, a new prestige, which was denoted by the giver taking a new name. The receivers, in turn, would at their own potlachs try to outdo the original giver and thus rise to their own new prestige. Thus in some ways what may in hindsight be seen as a rather quaint system of one-upmanship, was in reality a sort of stock exchange, a continual redistribution of the wealth derived from the diverse landscape in which they lived. The fascinating thing is that to feed all the guests at a big potlach required an immense amount of food, which could only be obtained by efficient hunting and fishing with first-class gear. So it was that subsistence, ritual and ceremony blended together into the perfection of a new set of lifestyles, which reaped the full potentialities of this very wet stretch of coast and its offshore islands.

We know that the ancestors of the coastal people came from Asia; we know not exactly when, but we do know that when the first white settlers came to those same shores not much more than two hundred years ago, some 100,000 Indians lived in a scatter of many hundreds of villages. They built gigantic ornate houses of wood, made canoes, carved totem poles and made ornate ceremonial head-dresses and regalia and kerf bent boxes. They spoke six different languages and dozens of dialects; and each group had its own special feature of custom, utility and ceremony, which fitted it to reap the benefits of its own particular locale. Whether, given time, those tribes we now call Tlinget, Tsimshian, Haida, Bella Coola, Kwagiutl, Nootka, Makah and Coast Salish would ever have become species we can only guess, for the tide of the twentieth century has all but swept them and their customs away.

Their rich lands are still there, for a contemporary Haida carver by the name of Bill Reid has written: 'Even today a stupid man could not starve on this coast, and today is not as it was.'

CHAPTER SEVEN

California says it with Flowers

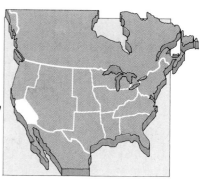

CALIFORNIA, the Golden State of Mind, the goal of the go-west trail, the richest of the rich, is the only place I know where you can buy a pair of 'his' and 'hers' helicopters.

Yet the extremes and extravagances of the Golden State in human terms, are equalled by its flora, which boasts the biggest, tallest, fastest-growing (though not the oldest) tree in the world. California is one of the most diverse workshops of evolution, set against the awesome reality of earthquakes. It also has a rich pre-history, of which little is known, and boasts Ultisols, Vertisols and Podsols about which we know a lot more, thanks to Hans Jenny.

California is not just one of the states which put the stars on the Union flag, it is the one that put the stars onto the cinema screens of the world. It is also a state of mind and so I am sure I will cause no dismay if, for the sake of my story, I bend its borders just a little. That little is enough to exclude the deserts to the east of the Sierra Nevada-Cascade Mountains axis and take in a hunk of Oregon, the north-east segment of the Baja California and adjacent Guadelupe Island.

The region within this new boundary enjoys a similar climate and hence vegetation: warm wet winters, and summers which are dry and hot enough to tan Hollywood's most vital statistics.

This area of approximately 324,000 square kilometres makes up the California Floristic Province, C.F.P. for short, the flora of which numbers at least 794 genera and 4,452 species of native vascular plants. That is more than the whole of central and north-east USA and adjacent Canada, a combined area ten times its size. What is even more staggering is that 6.3% of the genera and 47.7% of the species are endemic, which means that they are not naturally found beyond the bounds of the C.F.P. I defy anyone not to be bowled over by the sheer spectacle of the spring flowers of the Golden State.

Thanks to the media of film which had its own rapid evolution here in Hollywood (and may yet in the guise of cinema go to the wall of evolutionary extinction in competition with television), California has not only become linked with celluloid but also with San Andreas. Every popular account of Continental Drift mentions that the San Andreas Fault slices its way across the state north from the

Baja California, to disappear out to sea beneath the Golden Gate Bridge itself. It is a line of great geological activity with the ominous reality of earthquakes, and mountain ranges which are still in the making, as Mount St. Helens so spectacularly proved.

Early in the Tertiary period, a mere 60 million years ago, the area was very different. There were no high mountains and a tropical sea lapped against low hills which mark the contemporary position of the Sierra Nevada. That portion of California now west of the San Andreas Fault was about 320 kilometres to the south, a sub-tropical island which drifted in a series of complex movements north to take up the position it now occupies. The pressures set up by all this earth-moving (tectonic) activity has raised the Sierra Nevada and Coast Ranges in a series of complex upheavals combined with spectacular earthquakes and volcanic activity the rate and intensity of which has varied greatly.

At the same time the closing of the gap between North and South America with the formation of the Isthmus of Panama set the modern pattern of currents in the Pacific Ocean. Today cold waters well up along the Californian coast emphasising the difference between summer temperatures on land and in the sea. As cold air can hold much less water than warm air this effectively dries off the onshore winds. So it was that the Mediterranean climate developed about a million years ago, since when four ice ages have spread their cooling presence from mountain-top glaciers down into the valleys below.

Thus in terms of geography, climate, landscape and biology everything which greeted the first Californians as they made their way into this land of promise had all happened in the recent past, in fact it was still happening, for the post-glacial hypsithermal did not come to an end until some 4,000 years ago.

The contemporary picture of the C.F.P. is a landscape with high diversity and a climate which, though Mediterranean in character, ranges from that bordering on sub-tropical semi-desert in the south to Arctic alpine tundra above the tree line on the aptly named Sierra Nevada. The fascinating and most important feature is that despite all this change the climate of the bulk of the area has remained equable for the growth of plants. Throughout the past 60 million years it has neither become too hot and dry nor too cold and dry, the overall swing being from a warm wet temperate climate to a Mediterranean regime, the sort of difference you would experience travelling from Brittany down to the south of France, from whence the Mediterranean climate got its name.

To understand the effect this swing had on the vegetation of the C.F.P. all we need do is follow the fortunes of California's most famous plant, *Sequoiadendron giganteum*. But first a little background.

Around 50 million years ago forests dominated the land from British Columbia south to central California. They were very diverse: Alder, Birch, Cypress, Douglas Fir, Fir, Hawthorn, Maple, Pine, Poplar, Rhododendron, Spruce and many others formed a mixed canopy of both deciduous and evergreen trees. Beech, Chestnut, Elm, Holly, Hornbeam, Liquidamber and Sassafras, whose nearest allies are now found in America only in the Appalachians, were then abundant in the west, as were Dawn Redwood, Ginkgo, Keetleria and Pseudopanax whose nearest relatives are now found only in eastern Asia. We know this from fossil evidence thanks to the work of Dan Axelrod of the University of California. Together these trees were members of an immensely rich and widespread forest the remnants of which still demonstrate that the Eastern Connection of chapter 3 was much more than hillbilly moonshine.

However, even these diverse forests showed a marked and important change in composition at about the latitude of central Idaho. Both the conifers and hardwoods which grew south of this line were those which did not require very humid conditions for their successful growth. They were not plants of the true rain forest and few of their associates occurred further north, suggesting that even then drier conditions prevailed in that area. Most significant of all was the presence of a number of plants like Barberry, Mountain Mahogany, Oak, Sumac, and Ziziphus, all of which have the ability to control water loss during long periods of drought. It must however be noted that both these forest types were representative of a climate which enjoyed ample summer rain: a great mixture indicating changing conditions.

During the Tertiary period the climate slowly but surely got drier, especially important being a reduction in summer rain. The mesic (water-loving) forests in effect were forced to shift both coastwards and upwards where milder temperature and lower evaporation compensated for the reduction in summer rainfall. The trend continued over the next 40 million years and as it did so many of the east American and eastern Asian elements of the flora were lost from the west. Open grassland and steppe took the place of the forests in the north, and woodlands dominated by Juniper, Piñon Pines and Oak began to cover much of the Great Basin area which lay between the developing coastal mountains and the Rockies, parts of which were already in massive existence.

However I am jumping ahead, for by that time California's most famous plant was itself already coming into massive existence. The Giant Redwood or Big Tree is not the oldest, nor the tallest tree in the world. The first honour goes to a Bristlecone Pine which has already clocked up a very venerable 4,600 years; the second to a Coast Redwood measured at 112.77 metres in California, or a Peppermint Gum in Australia claimed at 182.87 metres, a difference which has already been well chewed over. It is not even undisputedly the largest tree by volume, for claims have been put in for the Giant Kauris of New Zealand. However the General Sherman Tree with a height of 83.02 metres and a wood volume of 1,416 cubic metres authenticated in 1931 is certainly the largest now alive and by anyone's reckoning deserves the name Big Tree. Its age has been estimated at 2,200 years, a mere stripling compared to the record Bristlecone and if correct would indicate that the Giant Redwood is also a contender for the title of the World's Fastest Growing Tree: an amazing 0.65 cubic metres of wood a year throughout a very long lifetime.

The ancestry of the Big Trees dates back at least 300 million years to the time when the first cone-bearing plants grew on earth and with their wind-borne pollen

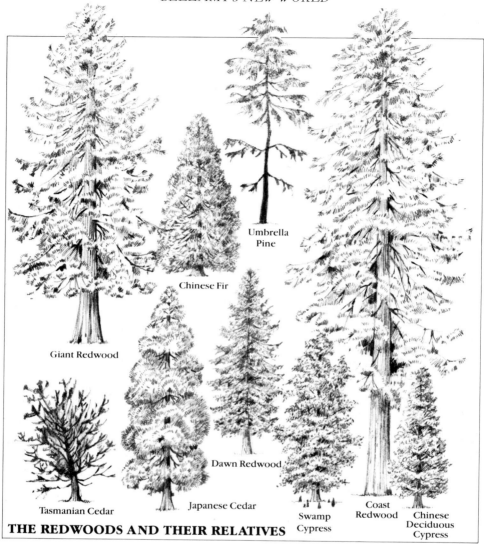

Umbrella
Pine

Chinese Fir

Giant Redwood

Dawn Redwood

Tasmanian Cedar

Japanese Cedar

Swamp
Cypress

Coast
Redwood

Chinese
Deciduous
Cypress

THE REDWOODS AND THEIR RELATIVES

and true seeds conferred upon the plant kingdom the freedom of all the continents which were only then getting it together in the form of Pangaea. Today the Redwood's relatives, the fourteen species which constitute the family Taxodiaceae, enjoy a rather strange disjunct distribution around the Pacific. They are: Bald Cypresses, three species, all natives of south and east USA and Mexico; Chinese Deciduous Cypress, one species found only in the Canton district of China; Chinese Firs, two species from China, Formosa and the nearby islands; Coast Redwood, one species native only in coastal California and Oregon; Dawn Redwood, one species native to a tiny area of eastern Szechwan and western Hupeh in central China; Japanese Cedar and Japanese Umbrella Pine, one species each, both natives of Japan; Taiwania, one species native to Taiwan and China; last but by no means least the Tasmanian Cedars, three species, all natives of Tasmania.

Fossil evidence dating from 125 million years ago shows that the Giant Redwood's three closest relatives, or at least their ancestors, were then widespread across the northern hemisphere, since which time their territories have all contracted. For example, 40 to 60 million years ago both Dawn Redwoods and the ancestral Redwoods themselves grew in abundance in the area of Yellowstone National Park. You don't even have to be an expert palaeontologist to appreciate the fact, for on the north-east slope of Amethyst Mountain you can see deposits that include the petrified remains of those and several other species of trees. Some of the stumps may stand where they grew and one measures 4.26 metres in circumference, a real big tree. A few million years later the Dawn Redwood no longer grew in America, its distribution gradually becoming restricted to a few relict sites in central China where it was discovered only in 1941 and brought to the attention of a war-weary world in 1944.

Fossils of a Redwood shoot and cone closely resemble those of today's tree.

To find a new tree, a living, no longer missing, missing link in the conifer line, was exciting enough, but when comparison of the fossils with the living tree showed that they were indeed members of a separate genus and not, as it had been thought, ancestors of the Redwoods or Swamp Cypresses, botanical interest grew and grew and everyone wanted to grow a specimen. So its discovery provided amenity forestry with one of its most beautiful trees.

Dawn Redwood is called *Metasequoia*; Coast Redwood, *Sequoia*; and our hero goes by the name of *Sequoiadendron*. The earliest fossils which undoubtedly belong to the latter genus were discovered at Trapper Creek in southern Idaho. They date from 2 million years ago and lie some 650 kilometres north-east of its present range. Other fossil localities are known, but all from sites on the eastern side of the Sierra Nevada and many are from contemporary semi-desert locations. From detailed study of these fossil floras Dan Axelrod has concluded that they were growing at an altitude of about 300 metres above sea level. The presence of Maple, Dogwood, Douglas Fir and Oak indicates a moist temperate climate with warm summers and cool winters, with 1,270 to 1,140 millimetres of rain well distributed throughout the year.

Today all the Giant Redwoods are found in 70 separate localities or groves scattered along a 420 kilometre stretch of the western flanks of the Sierra Nevada at altitudes between 1,370 and 2,285 metres. If Hannibal had trouble crossing the Alps with elephants, how did the Big Trees cross the Sierra Nevada all by themselves? Perhaps more to the point, why did they make the crossing? Why did the Big Trees go west?

The answer to the latter question lies in the drying climate caused at least in part by the most recent upsurge of the coastal mountains to their present snowy prominence of 3,550 metres. Even the impoverished forests could no longer grow in the developing semi-desert climate of the Great Basin which came to lie in a double rain shadow. The trees were thus forced either into extinction or to move west following, as it were, their own environment through mountain passes which had not then risen above their critical limit for survival. During the big push west many trees didn't make it, but the Giant Redwoods won through and the Coast Redwoods moved to take up their present strongholds along the coastal strip where sea mists distilled from those cold upwelling waters make up in part for lack of summer rain. Their way west was effectively being prepared by the developing Mediterranean climate which banished the Eastern Connection trees to other lands which still rejoice in summer rain, opening up the broad hectares of the C.F.P. for new colonisation.

It was the climatic changes of the last one million years which shaped the contemporary distribution of both the Coast and Giant Redwoods, the former to its misty coastal strip, the latter to its 70 sylvan eyries perched on the western flank of the Sierra. That the Big Trees once formed a more or less continuous belt along the mountains, their northern and upper limits determined by winter cold and their southern and lower limit by summer drought, is a matter of much more than pure conjecture. The developing valley glaciers and subsequent melt water scour must have restricted them to the higher ground between the main canyons, the drought stress of the hypsithermal finally squeezing them into their relict groves.

Studies have shown that water stress is most critical during the stages of germination and seedling development. For such a large tree, the Giant Redwood produces a small seed which contains little in the way of stored food. The seedling must therefore rise rapidly to self-sufficiency and for that to occur it needs light and water, both of which are likely to be in short supply, especially when the seedling is in competition with its overtowering parents. It is in this stage of life that a somewhat unlikely ecological factor comes to the aid of the potential forest giants.

Apart from summer drought, the other most important factor at work in any Mediterranean climate is fire. It has been proven that forest fires can speed the drying of mature Redwood cones and hence the release of their seeds. Seeds falling into the thick mat of leaf litter usually found beneath a Redwood may well germinate but will soon succumb to water stress because the raw litter holds very little

97

Page 99: The flowers of
the Prickle Poppy (*top*),
Globe Mallow (*centre*)
and Indian Paintbrush
add their own colours
to the spectacular
background of the
Grand Canyon.

Page 100: Springtime in
the Appalachians: as
well as Rhododendrons
(*left*) a mass of other
flowers such as (*top to
bottom*) Magnolia
grandiflora, Fringed
Polygala, Jack-in-the-
pulpit and Dogwood
clothes the hillsides.

A Willow by a stream in the prairie, which is also home to the Common Sunflower (*bottom left*) and Texas Bluebonnets.

Page 102: Man-made additions to the prairie landscape—a 'nodding donkey' oil pump and a rodeo.

Page 103: The Aleutian Islands harbour a wealth of bird species, including (*left, top to bottom*) Bald Eagle, Emperor Goose, Pintail Duck, Red-throated Diver, and (*right, top to bottom*) Crested Auklet, Stellar's Sea Eagle and Aleutian Canada Goose.

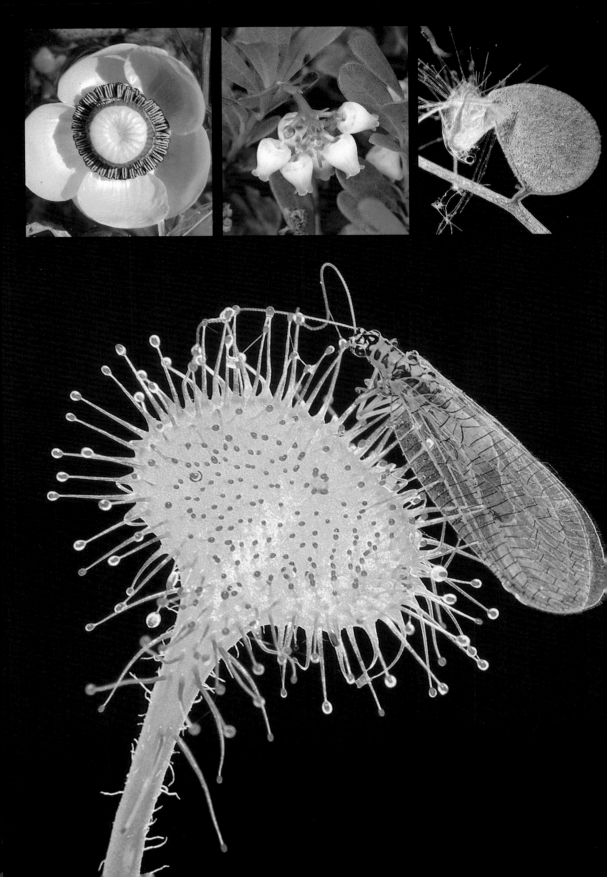

Page 104: Plants of the west-coast bogs: (*top, left to right*) Yellow Water Lily, Bearberry, and Lesser Bladderwort trapping Daphnia; and Round-leaved Sundew (*below*) with a Lacewing prey.

Page 105: (*Above*) Totem poles and Bald Eagle; close-ups of Haida totems (*below*) include a contemporary example by Bill Reid (*left*).

105

Page 106: In the Californian Chaparral (*right*) bloom many species which earn the state the name of Floristic Province: (*top to bottom*) Clarkia, Sand Verbena with Dune Primrose, Claytonia and the brilliant orange Poppy.

Page 107: A stand of massive Californian Redwoods reaches up through the mist to almost 100 metres.

Page 108: The Appalachians in the fall provide a dazzling spectacle of colour, and the bounty of the forests provided the early inhabitants with winter food stores of (*top to bottom*) Chestnuts, Bladdernuts and Persimmons.

Page 109: This scene could have been familiar to the Anasazi hunter-gatherers of a thousand years ago, as a family of Indians passes below the cliff dwellings of Montezuma Castle, Arizona.

Page 110: Amidst the grandeur of Chaco Canyon (*above*) the dwellings of Pueblo Bonito (*below*) were built in the 11th century A.D. by the people of the Anasazi culture.

Page 111: The marshlands of New England (*top*) where hay and cereals flourished (*left*) and cultivation of the Cranberry became a thriving industry (*right*).

The various habitats of
Manitoulin Island, such
as the limestones of
Cup and Saucer Bluff
(*top*), provide a home
for a wide range of
plants, including (*centre*)
the Manitoulin Wood
Lily, Common
Barberry, (*bottom*)
Shrubby Cinquefoil and
Mountain Avens.

in the way of available water. Fire burns off the dry litter allowing the seeds to fall onto mineral soil enriched by nutrients from the ash. There they can germinate, beneath the melting remnants of late spring snow, their roots penetrating quickly down into the organically rich mineral soil which holds sufficient water in available form. Fire will also have removed much of the ground cover, opening up the canopy and letting in more light, speeding both germination and ecesis.

Germination takes between 40 and 60 days and if all the conditions are right a few seeds will make it through to sapling stage. During their early life the saplings face many other problems: herbivores; competition from their parents and other plants; drought; disease; and of course the thing that helped them on their way, fire. However once an established tree has developed its thick fire-proof bark it reaches to maturity and beyond to the immortality of another seed, and remember it need only be one, well set upon its way to its own long life.

It is one of the greatest sylvan experiences on earth to stand amongst the Big Trees, each immense trunk like the column of some great cathedral church reaching up to the heavens which provide both the water and the energy of life. Likewise the hidden roots reach down to tap the source of mineral nutrients which are gradually being released to the living soil by the weathering of the parent material. Carbon, hydrogen, oxygen, nitrogen and a few other mineral elements borrowed temporarily from this special environment produce what is perhaps the most massive structure that has ever lived and will ever live upon this earth. Please, if you can, go and see them for yourself, stand in their awe-inspiring presence and think both of their past, their present, and their future, and when you do, don't stop at one, for each is different, part of the evolved diversity of a species which reaches back into all our pasts and on into all our futures.

If you can't make the trip to the C.F.P. then go and look at one of them growing in your own local park, for after their discovery in 1839 and announcement to the world in 1852 everyone, and that included most park superintendents, wanted to grow one of the Giant Trees. The first seeds arrived in the Old World on 28 August, 1853, where they were planted at Gourdie Hill near Perth in Scotland. These seeds and the many others which could not satiate the enormous demand even at the then Prince Consortly sum of £10 for a year-old seedling came from the Mother of the Forest Tree in Calaveras Grove. The bark was stripped from the bottom 36.6 metres of this giant and was reconstructed for the wonder not only of Queen Victoria but of all who came to the Great Exhibition at Crystal Palace in London. The mother tree was left naked, to die in the forest.

The fact that although restricted in nature they can be grown with ease in a whole range of environments is due to the care and watery attention they are given in captivity especially over that most critical seedling stage. The largest in Britain is 36.6 metres tall and may be seen at Leod Castle north of Inverness. The largest in Europe, where it is estimated that at least 10,000 are growing, is in the palace grounds at La Granja, Spain, where constant watering of the surrounding lawns since they were planted in 1880 has produced two magnificent trees, one 39.6 and the other 40.5 metres tall.

However, if you do make it to California and manage to tear yourself away from Disneyland and the bright lights of L.A. and San Francisco (and they are well worth seeing), I suggest that you also pay a visit to the forests of the Siskiyou Klamath region on the Oregon border, which receives more rain than any other part of the C.F.P. There, in a climate which is not unlike that which typified much of California way back in the early Caenozoic, you can have the experience of real forest diversity. Although a mere shadow of its former self of 50 million years ago, 16 to 18 species of conifer still vie for root space with many deciduous trees which now have their only western presence here. Sadler's Oak is found only here and in eastern America and Asia. White Spruce is the only American member of the Cembrae group whose other relations are in eastern Asia, the Alps and the Carpathians. Brewer's Weeping Spruce belongs to a section of its family all the other members of which are found only in Eurasia. These and many others occur there together, living proof that continental drift and environmental change are taking place and that the evolution of new taxa takes a long time. This represents a unique genetic resource reaching back into the past and an important part of all our futures, if only it is left intact.

So much for the northern, the mountain and the coastal forests, that section of the vegetation of the C.F.P. which is best called Arcto-Tertiary, for it came to California via the colder north during the past 60 million years.

What then of the vegetation which came from the warmer south? For fully one-third of non-desert California is, or rather was, covered with vegetation dominated by schlerophyllous plants. These typically have small leaves which being packed with mechanical (woody) tissue are hard to the touch and do not collapse and wilt when put under moisture stress. They in fact just shut up shop when the growing gets too dry and wait for rain. The best examples are the small-leaved evergreen oaks and Olives you see on your holidays to the European Mediterranean. Similar trees are there in force from central California southwards. In the northern section they are often intermixed with deciduous species and grade southwards through a scrubby maquis type of vegetation known hereabouts as Chaparral to semi-deserts which are outside the true boundaries of the C.F.P.

One of the best localities in which to get the feel of this north-south gradient is on Albany Hill, an isolated block of sandstone 90 metres high which faces San Francisco Bay. There aspect has telescoped the whole gradient into less than one kilometre. The northern flanks of the hill carry Live Oak forest rooted in deep dark soils and the southern slopes are covered with open grassland on shallow soils which are poor in organic matter. Between the two is the most marvellous zone of transition, an ecotone truncated into a few metres. Passing from north to south the oaks rapidly get smaller, their place being taken by grass as water stress and the likelihood of summer fires make their mark. Unfortunately Gum Trees, which are both drought and fire tolerant, planted in the early 1900s on real estate with a view of the harbour, are now making the natural pattern much harder to pick out. Travel north of San Francsisco and a similar hill will be entirely swathed in Oak woods while south the slopes of all the more arid parts will be covered with grasslands and in spring a riot of flowers.

Another very interesting type of vegetation found only along the coast is the Closed Pine forest. The Closed Pines constitute a group of eight species all of which can keep their mature female cones closed, some for up to thirty years, or until fire stimulates them to open up and shed their seeds. Three of the species, Monterey, Bishop and Nemoral Pine, are found in California, the other five in central and southern Mexico. It was within this now disjunct belt of coastal vegetation that Hans Jenny described his Pygmy Forest within which grows the endemic Bolander Pine. The Bolander is a subspecies of the Beach Pine, differing from it not only in its small stature but also in that its mature cones can remain closed on the tree for a number of years. There is good fossil evidence that the Closed Cone Pines evolved somewhere in the drier southern interior and came to the C.F.P., rafted north on that part of California which lies west of the San Andreas Fault. Subsequent climatic change and especially the recent hypsithermal have determined its localisation along the coast among the grasslands and the Chaparral.

The Chaparral is very rich in woody shrubs and with genera like the beautiful Ceanothus, represented by no less than 40 species, and Arctostaphylos (or as it is locally known Kinnikinnic) with 45, it can be very diverse. One species which is found throughout this truly Mediterranean area is the Chamise, a member of the Rose family with reduced needle-like leaves which like those of all the true Chaparral plants are ideal for conserving water in the summer.

The Chaparral habit, if such it may be called, is best looked upon as an extreme development of schlerophylly. Shrubs are often evergreen with small leaves which have a thick cuticle and sunken stomata. They are also typified by having very hard wood, deep root systems to tap the lower layers of soil in summer and the ability to re-sprout from their stumps after fire. These adaptations are not unique to the C.F.P. for Chaparral-type vegetation is abundant in areas which receive an abundance of summer rain. The fact is that Chaparral-type shrubs are common members of the plant communities from lowland rain-forest up to evergreen scrub above the true timberline, a case of convergent evolution in that the members of many different plant families show the same adaptations which allow them to tolerate water stress whenever it comes their way.

This south or Madrean element of the flora of the C.F.P. consists of 196 genera and 1,460 species, approximately 55% of the latter being annuals, which strongly suggests their rapid evolution since the development of the Mediterranean climate. As yet we do not know where this Madro-Tertiary flora developed, nor when it entered the C.F.P.. We do however know that by around 50 million years ago the Arcto-Tertiary and the Madro-Tertiary floras interfaced and intermingled with each other across a broad zone of gradual change, a gigantic ecotone. This must have interdigitated up the valleys of the developing mountains.

It is also safe to say that although the total flora would have been subject to the same broad vicissitudes of climatic and geological change, the effects would have been most keenly felt in this ecotone where contact, and hence competition, between the 'rival' floras would have been greatest. It is therefore likely that in this broad zone true equilibria between immigration and competitive exclusion would never have been reached and so free space would have always been available, space in which new immigrants and new products of creative evolution could find a home. The fact that continuing environmental change and upheaval would constantly bring new selection pressures to bear in the variety of micro-habitats available would speed the latter process.

That new immigrants have continued to enter the region is manifest, not only in the westward march of the Giant Redwoods but in the fact that 188 genera made up of 604 species, 241 of which are annuals, have entered the C.F.P. from the warmer south in the past two million years. Also that since the arrival of white settlers in May 1769 no less than 320 genera and 654 species, 57% of which are annuals, have become naturalised in the state.

It would be easy to say that the white man has destroyed enormous areas of vegetation and irrigated others. However the numbers given above only include those recent immigrants which have become truly naturalised. It does not include the 321 species which just hang on as agricultural and horticultural weeds. Each one of the 654 species has found a place for itself within the new semi-natural vegetation of the C.F.P. Of these new Californians only 110 came from the New World (50 from the USA, 23 from the tropics and 37 from South America). The other 554 all originated in the Old World and 480 of them came from the temperate and arid areas of Eurasia and North Africa—proof, or at least good circumboreal evidence, that the real weeds have evolved and spread across the earth with man, cash cropping in on his annual cycle of agriculture. In contrast very few Californian plants have so far gone the other way: the few include Amsinkia, Claytonia, Escholtzia, Lupins and Monkey Flowers.

What exactly is a weed? To a gardener or a farmer it is a plant growing in the wrong place. It is, however, much more than that, for even if helped along its way to that 'wrong' place, it must be able to make the journey and as travel is such a chancy business a plant must invest much of its living energy in the production of seeds. To do that it is a great advantage to become an annual, although therophyte would be a better term because many of them manage to squeeze more than one generation into a year.

The annual habit has one extra advantage, for it confers upon the plant the necessity of undergoing sexual reproduction every year and in the case of a really active therophyte even more often. At each fusion of the gametes (the eggs and the male nuclei from the pollen grain) recombination of the parental genetic material takes place and results in new genetic diversity and hence a broader spectrum of potential, ideal for a plant going out into a frontier or pioneer situation where change and disturbance is the order of the way ahead.

Imagine that a new genus, we will call it Clarkia, came into being somewhere in the region which was to become or act as a seed source for the Californian Floristic Province. Its method or origin need not concern us, but some chance series of mutations and recombinations within the ancestral line of the Evening Primrose family must have set this new taxon on its annual way to success. A plant able to make use of the spring rain to germinate, grow, flower, fruit, set seed and be back safe underground before the drought of summer will nip all but the best adapted perennials in the bud. A group of plants with those characteristics would be under enormous selective advantage in an area of developing Mediterranean climate.

Imagine all the different opportunities on offer: slopes facing north, south, east, west and all points in between; a great variety of rock types in various stages of soil formation, some almost devoid of vegetation, supporting only pioneer communities, others under grassland, scrub, wood or forest; all in a state of ecological change. In time, as the vegetation of each habitat within each landscape began to stabilise under the new climatic regime, some of the opportunities for annual plants would be taken up by populations of Clarkia, each individual population being the one at that time best fitted to the local environmental conditions, the diversity of environment, as it were, drawing out the variation inherent in the

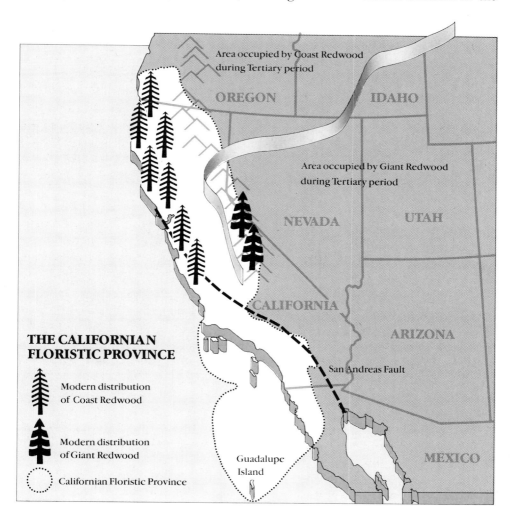

THE CALIFORNIAN
FLORISTIC PROVINCE

Modern distribution
of Coast Redwood

Modern distribution
of Giant Redwood

Californian Floristic Province

Area occupied by Coast Redwood
during Tertiary period

OREGON   IDAHO

Area occupied by Giant Redwood
during Tertiary period

NEVADA   UTAH

CALIFORNIA

ARIZONA

San Andreas Fault

Guadalupe
Island

MEXICO

genus, giving it fuller expression across the habitats available. During this process many of the populations would become distinct recognisable entities.

Imagine . . . But there is no need to imagine, it has already happened and the Clarkias are there to prove it, 43 species of annual plant, all except one of which grow within California and most of which grow nowhere else. What is more, many other genera have followed suit, producing new breeds of Californians.

It is therefore no coincidence that 55% of the native Madrean flora of the C.F.P. are annuals and that many of them are endemic, for they have evolved under the pressures and opportunities created by the fluctuating environment of this broad ecotone which still exists between the two main types of vegetation. Work on genera such as Clarkia, Cryptantha, Hesperolinon, Lasthenia, Mimulus, and Phacelia, all of which probably evolved within the C.F.P., is helping us to understand both the pattern and the process of creative evolution.

Thus we have it, the Californian Floristic Province, botanically one of the richest slabs of real estate in the world, a living museum with many paleoendemics (old endemics) relicts from the past, like the Giant and Coast Redwoods which have now been taken to many other parts of the world to enrich the landscapes and lifeways there. Likewise the many neoendemics (new endemics) show that it is also a vibrant workshop of evolution, a hotbed of (r)evolutionary change striving to make the most of all the opportunities on offer. Its 50 endemic genera and 2,151 endemic species are a unique living legacy from the past and an immensely important investment for the future of a world which is now dominated by one species: man.

So what of *Homo sapiens* in the C.F.P.? What of the first real Californians? It is a facinating question as to the immediate origins of the first settlers of the Golden West. There is no reason to doubt that their ancestors came via ancient Beringia but how did they enter the C.F.P.? If they came via the west coast route, were they descendants through and hence dependants on the Arcto-Tertiary flora? Did they come west across the mountains via the drying Great Basin? Or north-west along the route of the Madrean flora itself?

The west coast route was washed with the benefits of the sea and all it had to offer in the way of protein on the fin and within shell and carapace. The heavy forests of the coastal strip and mountain ranges would have provided game in plenty and in season but little in the way of land suitable for settlement, for even the best will in the New World backed by fire and the most advanced polished stone axes could have done little against the fireproof might of the Redwoods. Likewise the high mountains, many of which at first were capped, and the high passes filled with glacial ice would not have provided the best of routes inland. So man may have well stuck to the coast.

The more southerly mixed forest and especially the open woodland and savannahs dominated by Oaks of the Madrean zone certainly had more to offer both in the way of space and species for an inland way of life including as we know the big game which roamed the broad pastures of Wiltshire Boulevard and the dizzy heights of Beverly Hills around the Rancho La Brea.

It has been estimated that before the advent of the white settlers California supported some 150,000 people all of whom depended on a staple diet of acorns.

Acorn flour is both good to eat and full of nourishment, 21% fat, 5% protein and 62% carbohydrate, its fat content being much greater than that of either corn or wheat. There is, however, one problem, and that is that mixed amongst all the goodies is tannic acid, which makes the flour taste very bitter. Fortunately this is water soluble and can be removed by soaking. At some time in the past the local Californians developed millstones to aid the process and certain rocks in the state are pockmarked by hundreds of neat depressions worn by eons of acorn grinders. The acorn flour was then taken and placed in leaf-lined depressions made in sand along river banks. Hot water was then poured through the mass, rapidly leaching out the bitterness and leaving a tasty acorn cake which had the added advantage of storing and hence travelling well.

Millstone holes in which the original Californians ground their acorn flour.

This we know and much more, but what is lacking from this sunny Californian prehistoric scene are firm dates. Until many more of these artefacts have been studied in detail and accurately dated all we can do is guess as to the route which led man into this golden state of mind, and beyond.

CHAPTER EIGHT
Appalachian Fall

WHETHER the first humans encountered the bounty of the eastern forest from the north, along the central continental route, or the south-west, via the exotic extremes of California and the west-side strip, we will only know when many more sites have been excavated. We are, however, sure that the richness of the forests, and especially of their autumn harvest maintained by their Ultisols and Alfisols allowed two societies of hunter-gatherers to develop complex cultures and to take a major step toward civilisation.

As the summer draws to a close, the whole pace of life begins to slow, and plants and animals alike prepare for the long winter ahead. Leaves, fruits and seeds which had until then been focal centres in the life processes of the plants, the former as producers of energy, the latter as storers, no longer play that role and so begin to senesce and eventually to die.

The first indication that fall is just around the corner is when the green of the forest begins to wane and the first blush of autumn starts to tint the leaves. At first it is a mere hint of change, a patchiness edging along the veins, but as the now functionless system of photosynthesis breaks down the colours of the other pigments present, which until that moment have been masked by the all important chlorophyll green, show forth in all their vibrant glory.

Autumn is another Appalachian must, a pastiche of natural colour unsurpassed anywhere else on earth. There are not enough colours listed even in Oxford's biggest dictionary to express the subtle variety shown by each tree as it changes from its summer into its autumn garb. The whole forest covering those rolling hills and plains becomes a collage of millions upon multimillions of pendant palettes of colour as environmental change washes away first the greens, then the reds, oranges, yellows and browns. Each leaf, as if revelling in a job well done, puts on a final and most awe-inspiring show which will be remembered until spring breathes new green once more into those same branches.

Autumn is a time of great beauty and a time of plenty especially for those denizens of the soil who make their livelihoods from the products of decay. So too for the multitude of animals who live upon the bounty of the forest, for it provides them with an ample supply of nuts and other fruits and seeds to gorge into the fat of hibernation or store away for prudent use throughout the winter. The same was true for man.

When men arrived upon the scene the forests were there waiting, full of spring promise, summer game, autumn glory and with ample wood to burn and to construct a winter shelter to see them through safely into another year.

We know that the big-game hunters passed that way in force between 11,500 and 11,000 years ago for they left many Clovis points behind, clear across Kentucky, Tennessee and Ohio. Were they fluted like the later smaller Folsom points to fit into the cleft tip of a spear or dart? Or was it for some other reason? We can only guess but we know that they were used to hunt big game, both on the open plains and in the forest; some would add, to hunt it to extinction.

Whatever the actual reason for the mass extinction of these large mammals, the hunters were slowly but surely forced to rely on smaller game and the gathering of fish, shellfish and the edible parts of plants. Fish gathering may seem somewhat a misnomer but as the waters of the mighty Mississippi and its tributaries fell each summer, many fish would become land-locked in oxbows and other flood plain lakes and ponds from which they could be gathered with ease.

Likewise, the flood plains, though swampy, at certain time of year provided open highways throughout the forest and rich supplies of other types of food: waterfowl and their eggs; and many plants, the latter including Marsh Elder, Pinkweed, Blue Vervain, Goosefoot, Amaranths and Great Ragwort, which bear edible seeds in abundance, mainly in the later summer and autumn, American Lotus, Yellow Nut Grass, many Sedges and Duck Potato. Marsh Elder, Pinkweed and Blue Vervain were of most importance in the spring, and plants like Marchcress supplied edible greens throughout much of the year.

The list of plant products provided by the various types of forest is so many and so varied that it would become plain boring, although the diet derived therefrom could never have been called that.

The studies on ethnobotany, that is the plants which were used for food, medicine and construction by early man, are legion and indeed have recently come into vogue again because the modern giant drug companies have suddenly realised that at least some of their folk medicine was founded on fact.

One of the best studies to date relating to the ethnobotany of these woodlands is by April Allison Zawacki and Glenn Hausfater and has been published by the Illinois State Museum.

Starting early in the year—and remember after a long winter with all stored stocks running low, this could be the most critical time—the rising sap in woody plants was most important as a source of energy. From March through April this upward flow of dilute sugar from Sugar Maple, Box Elder, Silver Maple and Linden was tapped. Without the knowledge and the tapping technology this would have been a lean time in the forest, for little other food is in evidence.

Late April saw the first flush of edible greens: John's Cabbage, Lamb's Quarter, Butterfly Weed, and the stripping of young twigs of Slippery Elm and Green and Red Ash to obtain the young growing inner bark or cambium. The tubers of Pepperroot, Jack-in-the-Pulpit and Duck Potato were dug up and the swollen tuber-like bases of the new Sedge fronds, always there in time of need: hard work which uses up precious energy. Mid-June to mid-September is the time of ripening fruits and though the variety is enormous it must be remembered that succulence more often than not goes hand in hand with water content. Here are a few from the summer menu: Red Mulberry; May Apple; June and Dew Berries; Strawberry; Black, Common, Ground, and Choke Cherries; Wild Plum and Black Haw.

With the hard labour of summer past, the bounty of the autumn was in prospect. By adjusting your collecting altitude it was possible to extend the fruits of all seasons' labours by a few more days and in the absence of a method of preserving them, a few days were of immense importance.

The bounty of the autumn. Fruits there still were in plenty, Hackberry, Black Locust, Winter Grape, Persimmon and Hawthorn and nuts enough to go nuts about. I have already discussed Chestnuts, and even though they are now gone, their diminutive cousins, the Chikapin, are just as full of rich white meat. But were there ever really enough to support a large population of men whose ancestors had gorged themselves on the meat of Mammoth, Giant Bison, Ground Sloth, and even Giant Beavers the size of modern Grizzly Bears?

From basic ecological principles it can be argued that if the forest could support the big game animals which in turn supported carnivorous man, then with the big game gone the forest should be able to support even more men living more as herbivores. Although in essence that statement must be true, the equation unfortunately is not so simple. The demise of the megafauna must have had far-reaching effects on the structure of the forest, effects which may have changed its productivity for better or for worse.

However, using the status quo as set out in April Zuwacki and Glen Hausfater's paper, it seems safe to conclude that an average chunk of woodland real estate, which would include several forest types, would be productive enough, especially when it came to autumn. Each hectare of flood-plain forest could produce about 300 kilos of Hickory nuts, 300 of Black Oak, 30 of White Oak, 30 of Burr Oak, 10 of Red Oak acorns and 700 of Pecans. Likewise, each hectare of the forest which lies above the reach of floodwaters and in fact covered the vast majority of the land west of the Mississippi could produce 3,500 kilos of Hickory nuts, 750 of Black Oak, 300 of White Oak, 15 of Red Oak, 3 of Burr Oak acorns and 10 of Butternut. Then there were lesser amounts of Black Walnut, Kingnut, Swamp White Oak, Pignut and Shagbark Hickory, Mockernut and Bladdernut and many more, which could be eaten fresh, or stored away for winter. All this adds up to an amazing total of 1.3 and 4.5 tons respectively per hectare. No wonder they call it fall, not autumn hereabouts!

Whether the pre-chestnut blight forests were more productive than that or whether the productive place of that king of forest trees has been taken up by some of the others, we shall never know. What we do know for sure is that for more than 7,000 years, a gradually increasing population of hunter-gatherers evolved what is known as their Archaic Culture within these well-watered forests.

The promise of the easy pickings provided by the big game, and perhaps the excitement of the hunt, had kept the people on the move. This nomadic life had allowed them rapidly to colonise the Americas, both North and South.

Hunting the smaller game and gathering the richness of each season's crops was much more dependent on local knowledge, knowledge which could only be gained from a more sedentary way of life. A diverse stretch of terrain served by a winding river provided such a focal centre, within which perhaps two or three camps, each used at different times of the year, would be set up.

New methods of hunting were also needed and the atlatl or spear thrower came into being. It is a device which effectively lengthens the human arm and thus increases the velocity of the spear or dart, together with its projectile point, enabling the smaller, faster-moving animals to be killed. The hunters also polished stone and fashioned from the same material pots both for cooking and for storage. They also carved ornaments and beads from shell, indicating that this was not an all-work, no-play existence, but one in which life was rich enough to allow for leisure pursuits.

The storage pots are of immense importance, for adequate storage was the key to survival over periods of shortage. Stone cooking pots provided their own special problems, for how can they be heated enough to boil water without cracking? The answer was to fill them with water and drop in hot stones. If the latter cracked, the only problem was a gritty one, without the necessity of grinding out another bowl. Cooking made many of the products of the forest more digestible and wood, the main product of the forest, did the cooking. Black Locust and the Hickories produced wood which when it comes to heat production could approach or even surpass that of bituminous coal, with Butternut and Basswood only half as efficient. If my trial and error at my first scout camp is anything to go by, they soon would have learned to respect the difference. Carrying wood to do the cooking and stone pots to store and cook the food must have helped to temper their nomadic ancestry to a more static existence. Thus man in the diverse guises of the Eastern Archaic tradition, became part of the woodland ecosystem and lived, it would appear, in moderate peace and harmony with his woodland home, and so it remained for over 7,000 years.

Around 3,000 years ago, great changes began to take place, for the people, especially in the richer lowland forests, began to practise agriculture and with it a more ordered village life. It is suggested that the know-how came up-river, perhaps from Mexico or the south-west, for as we shall see both the necessity for and the potential to develop agriculture had already arisen there.

Using polished stone axes and fire, the rich bottom lands were cleared and a variety of crops were grown, at first probably in garden plots and later in larger fields. The crops included Jerusalem Artichoke; Sunflower; Sumpweed; and other rapidly-growing annuals, all developed from local stock. These together with the all-round abundance of the forest evidently produced a still richer way of life, with more leisure time manifested by the manufacture of a whole range of decorative goods. They made extraordinary pots which could be set on the fire to boil;

hence new freedom was conferred upon the cooks. There was time to fashion rings and beads and bracelets from native copper, imported from the north-west. Time to sit and think.

We know all this and much more about their technical skills, religious developments and other social structures because they buried their dead in furnished graves and tombs, which they covered with large mounds of earth. It is from within such mounds that the vast majority of their superb artefacts have been recovered.

They also built more complex effigy mounds in the shape of the birds and beasts which inhabited the forests with them. My favourites are a line of ten marching bears, with three attendant birds of prey which meander through the forest near McGregor, Georgia, and the Great Eagle with a wingspan of some 40 metres made out of white pebbles, also in Georgia. Perhaps strangest of all is the Great Serpent mound, all 1,254 metres of it, snaking its way through the woods near Jackson, Ohio. No artefacts have been found to date within its twisting form, but a burial mound close by has revealed artefacts attributable to the Adena culture.

The Great Serpent Mound, built by the early inhabitants of Ohio.

It would appear that subsistence agriculture and mound building were taken on as new practices by many of the diverse people who lived in the forests, a fact which linked them into what is now called the Woodland Tradition.

In time, two traits centering on Ohio became dominant within this tradition. The earlier Adena culture led the people to build conical mounds over a period of some thousand years, a period which ended abruptly around 2,000 years ago. Its place was taken by a culture which built round mounds with flatter tops and these, the Hopwellians, disappeared without trace in their turn, 700 years later.

They were evidently part of a trading network which stretched from the far north down to the sea. We know this because they made elaborate headdresses from copper, translucent drinking cups from sea shell, knives from black obsidian, graded and worked freshwater pearls into designs, and much more. Where they came from and why they disappeared are two of the great mysteries of America, mysteries which remain to be solved as more of the many thousands of mounds are excavated.

However, even at this inadequate state of knowledge we do know that neither of these advanced cultures ever became wholly agrarian. Agriculturalists of a kind they were, but much of their affluence was still derived from the natural affluence of the forests in which they lived.

They constituted a hunting, gathering, gardening society, which could build the Great Circle Earthworks at Newark, Ohio, a perfectly circular grassy embankment, with earthen walls up to four metres high or a little more, four hundred metres across. In the centre of the circle are four other mounds. Nearby is a square earthwork, and another of octagonal design, each impressive both in size and in geometry and each linked by corridors between perfectly parallel earthen walls. A similar corridor four kilometres long, extended almost straight down to the river. These are the trappings and achievements of a very advanced society.

The Great Circle Earthworks at Newark, 400 metres in diameter.

All this came and went by A.D. 700, at which time a new pollen grain was beginning to blow on the woodland air.

Chapter Nine
US 'Amaizing'

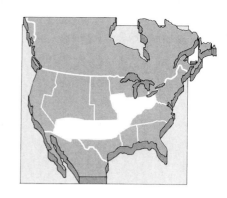

THE new breed of plant which was to change everything for man came into chance existence somewhere to the south of the USA. Its name: maize. Its fate lay in the hands of man; its destiny was to be grown on all types of soil in the New World and far beyond, creating cultures, civilisations—and problems—of ever-increasing magnitude.

The pollen grain in question is monoporate, which means it has only one pore; it is roughly the shape of an American football, though of course much smaller. It belonged to an annual grass, which on completion of successful flowering produced a head or cob covered with fruits. Its Latin name, *Zea mays*, known today to multi-millions across the world as Corn-on-the-Cob, Sweet Corn, or just plain Corn.

The cobs of cultivated Sweet Corn (*left*) may have originated from perennial Teosinte, shown here with Rafael Guzmán, Mexican discoverer of this population at 2,300 m.

In nature it is a grass with a distinct problem, for it has no method for dispersing its seeds. There is no natural method for getting the corn off the cob! A ripe cob left on the plant will eventually fall to the ground and germinate en masse. The seedlings will thus die in competition with each other—and only with luck will one come to the fruition of flowering and fruiting.

So it is safe to conclude that, without constant help from man, corn as we know it would become extinct in a very short time. For similar reasons it is argued that corn must have come into being with the help of agricultural man somewhere in the Americas and, as we shall see, it is equally true to say that in America agricultural man came into being with the help of this botanical monstrosity.

The burning question thus appears to be, not which came first, but whether man the gatherer simply found the species Wild Corn and improved on it by constant selection or whether, by selecting and growing corn's ancestors side by side, he played an active part in the evolution of the species *Zea mays*.

Whichever way it was, the subsequent selection and development of the crop by the first farmers of the American continents ranks amongst man's greatest achievements.

To date the oldest known corn cob was found in St. Marco's cave, Tehuacan valley, Mexico and dates from 7,000 years ago. It is about 20 mm long and, though small, is undoubtedly a corn cob which bore a mass of loosely packed seeds. One group argue that these are the cobs of Wild Corn, others that it was a part-way stage in the evolution of the species. The proponents of the second theory look to a grass called Teosinte, which still grows wild in the highlands of Mexico, as a possible ancestor of Corn.

The second part of their claim is backed by the fact that although *Zea mays* (Corn) and *Zea mexicana* (Teosinte) are morphologically worlds apart (in appearance they look quite unrelated), genetically they are so close that they can be hybridised with ease. In fact, hybrids are quite commonly found in nature where corn is grown within the habitat of Teosinte and, what is more, the hybrids show a great deal of fertility. It must be remembered that the hallmark of a true species is that it cannot be crossed with another species to produce fertile offspring. For this reason, Dr Hugh Iltis of the University of Wisconsin and a major proponent of the Teosinte theory calls the wild grass *Zea mays* ssp. *mexicana*.

In order to prove his point, Iltis continues to do much interdisciplinary research into the problem, research which started out with a massive experiment aimed at determining the number of major genetical differences which exist between what he recognises as two subspecies. The work was based on Mendel's Laws of Heredity, which, although first enunciated in 1865, even today form a cornerstone of modern molecular genetics and plant breeding.

Mendel's laws indicate that if two interfertile individuals which differ by only one gene are hybridised, then each of the original parent types will reappear in the second generation with the statistical frequency of one in four. Two independent gene differences and the reappearance ratio becomes one in sixteen and with ten such differences almost one in a million.

Taking a long shot, and cutting the odds by choosing Chapalote, a primitive Mexican Corn, and Chalco, the Mexican Teosinte which looks most like corn, Iltis grew 50,000 second generation plants. After an immense amount of painstaking work over a number of seasons, the frequency of good parental types was one in 500, pointing to only some five major genetic differences between the two. There is thus little doubt that they are interrelated quite closely, but whether by direct ancestry or subsequent inbreeding still waits to be established.

I could go on, for I think it is one of the most interesting pieces of botanical detective work of recent years and the arguments and counter-arguments have spiced every International Botanical Congress since I first attended one as an undergraduate. I must add that, however corn came into existence between 5000 B.C. and A.D. 1492 when Christopher Columbus first saw 'mahiz' (that which sustains life) being grown by the Indians on the island that is now called Cuba, the descendants of those hunter-gatherers which poured across the Bering land bridge had done their best to reap the full benefits of the New World. In so doing they had developed successful agricultural systems in a whole range of environments from the mouth of the St Lawrence River in Canada, clear down to central Chile, all based upon two to three hundred highly selected and interbred varieties of Corn. These included all the ancestors of the modern varieties that today form the basis of a multibillion dollar industry—red, blue, yellow, sweet, field, flint, flour, pod and popcorn. There is also little doubt that in those areas of tropical and subtropical America which rejoice in short days, it hybridised with a number of Teosintes including *Zea mexicana* and *Zea luxurians*. *Teosinte* is derived from an Aztec word which means 'God's Ear of Corn' and to this day in many parts of Mexico it is known as 'Mother of Maize', facts which may in part back up Iltis' claims.

Some of the many varieties of selected and hybridised corn which today grow in a whole range of environments and feed the hungry world.

However the original corn plant came into existence, it was much more desirable to man than Teosinte or any of the other grasses which have been suggested as amongst its ancestors. It was also much more adaptable and could be grown in latitudes where the long days of summer precluded the flowering of Teosinte. The tassel-like male flowers of Corn produce pollen which, like that of all grasses, is blown on the wind. What is more, it can readily be identified because it is larger than any of its proposed progenitors. Wherever it is found in sub-fossil form it signifies that it was blown there on the winds which were to change the face of America.

As the last ice age came to its close the whole climate of America and especially of the South-West gradually became drier. The melting of the mountain glaciers of the Rockies and the High Sierras filled the deepest parts of the great basin which lay between the two with water. The lakes overflowed with fish and became a focal centre for game, both big and small, and the hunters that followed their trails. Lying in a rain shadow of both mountain ranges the lakes, once deprived of their glacial inflows, began to shrink, evaporation being far in excess of the meagre precipitation. The salts dissolved within the waters eventually rendering them unfit to support anything but the most highly adapted forms of life. At length they dried up to produce the famous Bonneville Salt Flats and leave remnants like Mono Lake and the Great Salt Lake itself. The big game and their hunters came and went, their place taken by men with the will and skill to hunt Antelope and smaller game and fish in the dwindling lakes.

Perhaps more than any other people then living in the New World their lifeways became moulded by the vagaries of a hot, drying climate. Survival forced them to hunt ever smaller game, each kill taking much more effort and skill with the atlatl. In time even insects became part and parcel of their daily diet, as they were forced more and more away from hunting to gather the fruits, and especially the seeds, of the expanding semi-deserts, each one in its season. For this they needed baskets, both to aid collection and for storage, and it is here we find the first baskets ever to be used in the New World. Likewise, by 8000 B.C. they were using special tools to grind the seeds into flour, absolute proof that the technology developed on the spot and was not brought in by arrivals from other lands, for they predated the first millstones of the Old World by almost 2,000 years. It would appear that, in those areas with the harshest climate like the Great Basin and the surrounding semi-deserts, this remained the local life pattern until recent times.

However, in the Four Corners region, which encompasses the States of Utah, Colorado, Arizona and New Mexico, the winds of environmental change were also blowing but here they were to give rise to the most amazing cultural revolution.

Perhaps it was that the upland climates of the region were damp enough to support open forests of Douglas Fir, Rocky Mountain Juniper and Limber Pine, all of which support game in abundance; also that by moving both in altitude and in latitude, the local hunter-gatherers could make use of a number of zones of climate and hence vegetation, thus avoiding the dire effects of any prolonged period of drought. Or perhaps it was simply that it was close to that important spot somewhere in middle America where a gatherer first realised the potential of an annual grass and took the first step towards an agriculture which would fill the bellies of generations of Americans with corn and the air with its distinctive pollen.

One problem with pollen is that it gets everywhere and so a few grains found in sub-fossil form can give a distorted idea of the importance of a plant in any particular place. Great care must therefore be exercised when interpreting the pollen record and only when many adjacent sites tell the same sequential story can its detail be relied upon.

Unfortunately, the Four Corners region does not abound in lakes and swamps in which a sequential pollen record could be preserved. The majority of the finds have been in gravels and silts deposited and hence sorted by water action; and the record of the vegetational change is far from complete and, in most cases, its interpretation is very hazy.

There is, however, enough firm evidence to say that, during the last glaciation, that is between 35,000 and 12,000 years ago, the climate was wetter and large areas which now bear semi-desert vegetation were covered with open forest of Limber Pine, Spruce and Douglas Fir. During this time the upper tree line was lowered so much that treeless alpine tundra dominated by Artemesia occupied mountain summits which rose above 2,700 metres even in the north of New Mexico. The zones dominated by trees were thus both telescoped and pushed downslope to form lowland corridors through which Bristlecone Pine, Limber Pine and Spruce migrated to their now disjunct locations on the summits of mountains in the southwest. Then between 12,000 and 10,000 years ago, as the ice age came finally to a close, there was a rapid transition to the modern types of vegetation.

In the light of the inaccuracies inherent in pollen analysis, especially when the pollen is poorly preserved, the emphasis lay in finding more tangible macroscopic remains which would present an accurate picture of the make-up of the past vegetation of certain localities. They were found, once the habits of a certain small mammal, the Packrat, were understood in detail.

Packrat middens provide evidence of vegetation types dating back to 8600 B.C.

It is now known that Packrats gather plant materials from within 30 metres of their dens and that the remains accumulate as waste dumps or middens. In caves or within the protection of overhanging rocks such middens can be preserved for very long periods of time. Just what the archaeologists ordered; for being organic, they can be accurately dated using the radio carbon technique and so the middens containing organic remains can be placed in an ordered time scale.

In a setting which is now well within an area of semi-desert, the oldest middens, which date from 8600 to 7600 B.C., are dominated by the remains of Limber Pine, Rocky Mountain Juniper, Douglas Fir, with a little Spruce thrown in for good measure. Ponderosa Pine, one-seeded Juniper and Piñon or Edible Pine make a later appearance only in younger middens. After 3550 B.C. all the first five have

THE THREE FACES OF THE FOUR CORNERS
Cultural advance and vegetational change

UTAH

Anasazi
Albuquerque
Flagstaff

ARIZONA

Phoenix
NEW MEXICO

Hohokam
Yuma • Tucson

Mogollon
MEXICO

1. Basketmaker

2. Hohokam

3. Anasazi

disappeared, their place of woody dominance taken by the last two intermixed with the majority of the woody species found in the contemporary semi-desert vegetation. Finally Piñon Pine bows out of the Packrat life-style and out of the picture, about five hundred years ago when the whole area had gone over to a vegetation dominated by Yucca, Prickly Pear Cactus, with the odd One-seeded Juniper surviving in the most favoured spots. The youngest midden studied dates from 460 years ago and contains 87% of the perennial species of plant which are still found growing within 30 metres of the site; Four-winged Salt Bush, Cliffrose and Mormon Tea are the co-dominants, as they are today. So it would seem safe to conclude that the open forest which dominated the area throughout at least the terminal phases of the ice age was gradually replaced by Piñon and Ponderosa Pine as the local climate became warmer and drier. Regeneration of the latter is controlled by a good seed year coinciding with a wet summer, for once established the seedlings of this hardy plant can withstand drought and hence can grow at lower, drier altitudes than the other pines. Likewise, as the species is tolerant of fire, the summer monsoon climate, with dry springs followed by thunderstorms and wild fires, would favour its presence.

The work quoted above was carried out by Julia T. Betancourt and Thomas R. Van Devender of the University of Arizona Tucson in the area of the Chaco Canyon in New Mexico, where one of the man-made wonders of the world was discovered in 1840.

Its name, Pueblo Bonito; and it was built between the years A.D. 1000 and 1150, ending up with 800 living rooms and Kivas (probably religious meeting places) parts of which stood at least four storeys high. Built of shaped masonry, it was the work of skilled artisans, for the walls, though massive to support the vertical height, had their outer faces finished with alternating layers of thick and thin slabs set in various different styles. What is more, the roofs of both the houses and Kivas were supported by massive logs of Ponderosa Pine. It is estimated that over 100,000 trees were used just to build this and the other smaller apartment houses in the Chaco system.

Going by the Packrat evidence, the pine trunks must have been imported from quite a distance away from the building site. Roads ending in steps cut into the living tock of the canyon side indicate the route, but not the method, of transport.

This is just one of many such centres of habitation and activity of an advanced culture, which has come to be called the Anasazi; a culture—no, a civilisation—which developed across much of the Four Corners region. That the local environments were changed by its activities, and especially by its destruction of woodland both for building and for fuel, goes almost without saying. That it caused the final destruction of the forest and hence hastened soil erosion is as difficult to prove as it is to deny.

The logs used in the various constructions tell another story in intimate detail, for growing in what must have been a somewhat marginal environment with a fluctuating climate, their annual growth rings have recorded in their widths the good and bad years in intimate detail.

It was realised way back in 1929 that the environmental detail recorded in trees still growing and those present as logs in structures of recorded age could be compared, and thus linked with those from undated prehistoric sites. This overlapping of tree-ring data from older and older sites thus provides a continuous record of climate and hence an almanac against which to date past events with great accuracy. So accurate is the record that in one building, the Chetro Ketl near Pueblo Bonito, it was determined that several logs were cut in the winter of 1039–1040 and that others were cut in the following spring.

Using such accurate methods it is possible to date all the phases of cultural development of the Anasazi between 100 B.C. until the historical period began in earnest when in 1598 soldiers, missionaries and settlers from Spain started to change a way of life which had evolved to great perfection over a period of 1,600 years.

Throughout that time the natural environment had continued to control their way of life and that of other groups which lived in the areas round about. There is strong evidence that the years A.D. 600 and A.D. 1150 were good years and, in fact, were focal points in periods of higher summer rainfall and hence plenty. The growth rings on the Pine were wide and the Anasazi expanded both in population and ideas. It was around these periods that more lowland areas, which had been too dry or at least too marginal for survival in the past, were developed and occupied, only to be abandoned during the ensuing dry periods, which the tree rings show to be centred on A.D. 875 and 1425. It is tempting to conclude that the regional climate is undergoing a major cyclical variation with a 500-year periodicity; only time, lots more time, will tell if this surmise is correct.

Volcanic eruptions were not unknown in the area and added their own special problems to any way of life. In 1065 (the year before a very important event took place in British history) somewhere just to the south of the main Anasazi region a volcano erupted, producing what is now known as the Sunset Crater. The area was occupied by a group called the Sinagua ('sin agua' in Spanish means 'without water'), who, though less developed, traded with the Anasazi for decorated pottery. The eruption lasted for many months and greatly inconvenienced the locals, but those who stayed on found that the rain of cinders had so improved the water-holding capacity of their local soils that their agriculture began to show a marked improvement—so much so that they were able to adopt many of the Anasazi ideas and, from A.D. 1100 on, built masonry apartment houses. The conclusion, rightly or wrongly, must be that the soil improvement tipped the balance, enhancing the potential of their environment sufficiently to allow them to cash in on the best of the local ideas and cultures which had developed in more favourable environments round about. The close link between environmental limitation and man's aspirations is thus made ultra clear.

To a people dependent on the gathering of food or growing of crops, any factor which would swing the productive balance in one way or another must be of great importance. The line between starvation, subsistence and a little to spare is a narrow but very important one. For that little to spare, multiplied by many pairs of gather-

Sunset Crater Volcano, Arizona, formed by an eruption of A.D. 1065.

ing or agricultural hands, provides enough time for specialist developments within the community at large.

It was in all probability this changing marginality of environment that nurtured both the need for, and the possibility of, social change, which included the development and adaptation of agrarian practices to meet local needs and conditions. Once developed, the sky must have appeared to be the limit, unless the environment changed, or was changed, for the worse.

In the semi-deserts to the south-west of the main Anasazi region lived another group called the Hohokam. It would appear that they came north out of Mexico some 300 years B.C. to settle reasonably fertile lands along the Salt and Gila rivers. They not only brought with them crops and the techniques with which to grow them, but also maintained trade links with the developing cultures of the south. They supplied them with turquoise in return for Macaws from the forests and jewellery made from sea shells, and new ideas must have flowed in both directions.

So it was that the Hohokam began to develop what turned out to be massive systems of irrigation, which so improved their agricultural potential that they rapidly became the most urban of the prehistoric south-westerners.

The Anasazi were not slow to learn from their neighbours and they, in turn, began to irrigate their fields in the 10th and 11th centuries A.D. Up till that time they had used what had by then become an age-old method which maximised the chances of a good crop in such a chancy environment. Small, almost garden-sized, plots were planted with a mixture of Corn, Beans and Squash.

The Corn was staple in their diet, the 'flour power' which fuelled their aspirations—and the first record of it in the Anasazi region is from Bat Cave in New Mexico and dates from 3000 B.C. (although the date is hotly debated). It was a primitive Pod Pop Corn which, as its name implies, had its fruit enclosed in a little case or pod which caused it to pop when heated in the right way.

Pole Beans of the genus *Phaseolus* were grown to climb the corn stalk, where they flowered and bore fuit. Very early on the farmers must have realised the importance of this crop, both in their diet and to their soils. We now know the reason, for beans are rich in high-grade vegetable protein and the nodules on their roots are infested with bacteria which fix nitrogen from the air and turn it into nitrate fertiliser.

The Tepary Bean grows naturally in north-western meso-America and this may well have been the first to be domesticated by the ancestors of the Anasazi. However, Runner Bean was being grown in Mexico in 7000 B.C. and the Common Bean by 5000 B.C. While the Lima Bean was in use in South America, its native home, long before it was grown in Mexico in 500 B.C.

The Squashes, genus *Cucurbita*, have an equally interesting origin. They were grown as a ground-cover crop which helped to keep out the weeds and to provide good kindling for cooking the food, and also produced a good organic mulch for the soils in which they were grown. Pumpkins and Summer Squashes had been domesticated in Mexico by 7000 B.C. and the Cushaw Squash some 2,000 years later. The Walnut Squash and others followed on, new crops to get the best out of each plot and each growing season and new tastes to whet appetites for more.

It is of interest that the only main crop plant to originate within what is now the United States of America is the Sunflower, and although it was never an important part of the polyculture, it was and still is a rich source of oil.

It is clear that, once agriculture was an established practice, there was a considerable exchange of species and varieties from one area to another, and with them would have gone a detailed understanding of their requirements and hence their physiology. However, the evolving agriculturalists had to learn by trial and error the potentialities and problems of each crop and each soil type in their area.

The process of soil formation goes on very slowly in an arid climate and so deep, well-structured soils are a rarity. However, sandy soils and rocky soils with deep cracks and fissures filled with finer material will be wetted to greater depths by rain—greater, that is, when compared with soils rich in clay, which will soak up and hold the water closer to the surface. Between rains the latter soils dry by evaporation much more rapidly than the former and hence will best support only quick-growing crops.

Our agriculturalists must also have come to realise, though not in so many words, that productivity of dry matter by any crop closely correlates with the amount of water available to and hence transpired by the plants. That is, of course, provided that the soil is fertile. So irrigation could greatly improve the crop yield and, what is more, rapidly deplete the soils of their mineral nutrients. Corn, even when grown in watered polyculture, obeys the maxim that 'there ain't no such thing as a free lunch' (Tanstaafl), for though it provides much it demands much from the soil.

In an environment in which evapotranspiration is always going to be in excess of rainfall, without sufficient irrigation at all times unwanted salts of sodium and calcium will accumulate in the soil, causing problems of long-term fertility.

Want, the mother of necessity, backed by trial and error in the adoption of new and improved crops and technologies which came their way, bolstered their population and, with more hands to be put to other work, their arts, crafts and culture rose to new heights of expression.

All things, good and bad, must eventually come to an end and the Anasazi culture was no exception. As the tree resource was depleted, the soils were opened up to erosion by wind and water and much fertile topsoil was lost as flash floods took their toll of life and soil within the valley. The increasing population and their demands for more affluent lifestyles, based on higher yielding and more demanding varieties of crops, increased the pressure on already depleted soils, which became alkaline and fell into disuse.

In time, the still successful sites became oases in a rapidly expanding semi-desert, oases which not only attracted the locals but became the envy of marauding bands of others who knew little about farming, but were still expert hunters or warriors. The resiting of Anasazi villages in canyons and cave sites which were difficult of access could be in response to such attacks or simply that they offered perfect sites for construction close to areas which could be effectively irrigated.

The search is still on for the answers amongst the ruins of great apartment houses, which rank both in beauty and in age with any rose-red city. Just imagine a jewel-like dwelling fashioned from pink Navajo sandstone set in the sheer walls of a canyon! No, don't imagine, go and see for yourselves: Chaco, Canyon de Chelly, Keet Seel, Mesa Verde and many more are not just names from the past, they are sites of exploration and wonder which will always hold a special fascination for me and an importance to the world of human kind.

Hot dry winds signalled the drought of 1276. Those trees that were left recorded the fact, and fewer grains of corn pollen blew into the Packrat middens. The great farming communities were coming to an end in the north as the bulk of Anasazi endeavour moved south, leaving first the San Juan and eventually the Little Colorado river basins abandoned to Mesquite and Prickly Pear.

The civilisation come to an end but the technology of growing Corn, Beans and Squash, did not, for it was taken up and carried on by those who eventually came to take their place. It also spread across the length and breadth of the New World; in fact, to anywhere in which Corn could be made to grow, flower and fruit.

Yes, even back in the forested east. The end of the Woodland tradition signalled by the disappearance of the Hopwellians, was followed by a new tradition of agriculturalists called Mississippian, who took the art of mound building to even greater extremes, and we know they did it with the help of Corn. The greatest mound of all lay at the heart of a city called Cahokia which, until Philadelphia reached the 40,000 mark early in the nineteenth century, was the largest city in the area which is now the United States. Cahokia now lies within the sprawling suburbs of St Louis, and its excavation is going on apace.

The lifestyle of these Mississippians centred on small farm plots which provided the basic food requirements for a rapidly expanding population. Corn, of many productive varieties and increasing all the time, was backed up, both in mono- and polyculture, by Beans, Squash, Pumpkin, Sunflower, Jerusalem Artichoke and many other less important crops. Gathering of all the natural products the dwindling forests still had to offer—fruits, nuts, tubers, freshwater mussels—and the hunting of fish, wildfowl and small game were still carried on, but became of less consequence.

The increase in food produced, measured in terms of time and effort invested, became an important consideration as potters, basket makers, coppersmiths, bead makers and workers in flint and stone further perfected their crafts. To satisfy the need for raw materials, merchant traders travelled from the coasts of the Gulf and the Atlantic with conch and other sea shells, to Wisconsin, Oklahoma and beyond for high-grade flint, to the Great Lakes for copper and to the flanks of the Smoky Mountains for mica.

At home, they lived in square houses built of logs covered with mats or thatch. Many of these houses were grouped in planned clusters along avenues and around plazas, and others were set amongst the thousands of hectares of farmland which supplied the food, or along the roads which radiated out to the many satellite towns and beyond.

Dominating the great city and the lives of the Cahokians was the largest of all the mounds: a giant platform, its base covering no less than 6 hectares, its four terraces rising to a height of 30 metres. It was built in stages during the period from 900 to A.D. 1200 and was topped by a large wooden 'temple' and/or ruler's residence, 30 metres long, 15 wide and possibly as much as 20 metres high. Other lesser buildings occupied the terraces and other platform mounds in the vicinity.

What was it that made the people carry millions upon millions of basket-loads of soil up to build this and the many more mounds (there are well over a hundred others in Cahokia alone)? It could be argued that the problem of idle hands was forestalled by clever minds finding them work with religious significance to do— and that the fears of sacrificial retribution by the Gods kept them in earth-moving line. It could also be argued that, for a community which was totally dependent on the environment for their wealth and (what is more) knew it, religion arose out of that dependence and that the mounds were built as a way of reaching up to the unknown that had given them dominion over all the products of the forest. Perhaps it was a mixture of the two: who knows? But, as the excavations continue on apace, we are beginning to learn more about their customs. Cahokia is the most amazing place. It is the only place I know in the world where a local community is being moved out by the state, lock, stock and bungalow, to make way for the archaeologists. It is fortunate that the modern Cahokians live in clapboard and brick houses which can be put on wheels and towed away. The authorities of A.D. 1200 would have found it much more difficult, even with twentieth-century technology, to move the dwellings of the Mississippians.

The significance of 'Woodhenge', the remains of four circles, the largest being 150 metres in diameter, each marked out with large wooden posts, is still a focal point of discussion. The wood used was Red Cedar, a tree which pollen analysis shows was once abundant in the area and which is noted for its tall straight growth. Warren L. Wittry, who has not only doodled with the problem for many hours on graph paper, but has sat through dawn and sunset upon a seat set up where a central observation post once stood, is sure that it was a celestial observatory.

The post holes which mark the position of the sun at the winter and summer solstice were probably the most important to an agrarian people. Solstice means the time when the sun stands still, that is, stops its daily progression to the north or south and begins to return to light the skies, warm the seasons and charge the leaves of Sunflower, Squash, Beans and, above all, Corn with the energy only green plants can supply.

These were times of rejoicing and religious ceremony—and any person who could sit on the central post and so predict the exact day on which it would happen had power indeed.

Corn, as it had been developed over the millennia by selection and opportunistic breeding, had come to be the most efficient converter of solar into chemical energy present in the New World. Likewise, the efficiency with which man could control his destiny through competent agriculture also rose to new heights.

WHO DISCOVERED AMERICA?
Areas occupied by Indian tribes before colonisation by white man

But was it too efficient? For during the Mississippian period other great changes began to take place. Each township (and that included the great city of Cahokia) was furnished with a stout defensive wall of wood, which was replaced again and again as the old one fell into disrepair.

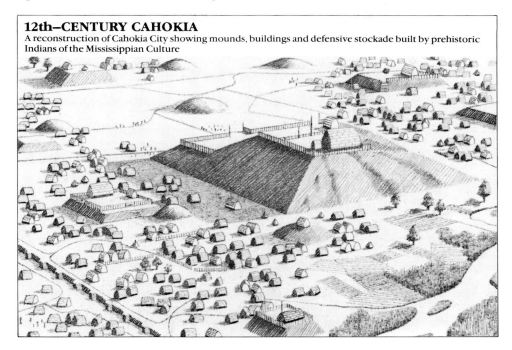

12th–CENTURY CAHOKIA
A reconstruction of Cahokia City showing mounds, buildings and defensive stockade built by prehistoric Indians of the Mississippian Culture

Was it that these rich oases of agriculture, within their sylvan setting, were just too much temptation, and so marauding bands from round and about and further afield began to demand their share? Was it internal strife? Did the workers who were told to carry soil and build the mounds still higher, seek elevation of another type within their own society? Again we do not know, there are no records to read, for this all happened in the pre-history of that 'amazing' place, America.

There is no evidence of catastrophic war nor of massive pestilence, or at least none has yet come to light; but by the year A.D. 1500 the culture was no more and the mounds were left silent to the creative depredation of forest regeneration.

The lands were not left empty for long and neither were those of the Anasazi or any of the other lesser cultures for when the Europeans came to the New World, they didn't find an empty land but one overflowing with a diversity of peoples: some 287 principal tribes of Indians in all, hunting, gathering, fishing and many still planting Corn, Beans and Squash; each tribe making the most of everything their particular bit of real estate had to offer. It is almost as if the winds of a change called civilisation had blown, held their sway and then passed on, leaving the real survivors, those descendants of the people of Beringia who perhaps found satisfaction in a simpler way of life which eventually became their saving grace.

CHAPTER TEN

Hay Days

THE first white settlers who came into the east made their homes amongst the bogs and forests developed on the Podsols and Ultisols of the coastal strip. Without the active help of the Indians with their local knowledge, they would not have been able to survive the first harsh seasons. They were quick to learn and soon were growing corn, beans and squash—the crops which had served America so well throughout its pre-history. They also used the local cranberries to add vitamins to their diet and flavour to their Thanksgiving celebrations.

From this firm foothold in the New World, they set out to colonise the continent, and like the Indians who had gone before them, made special use of special skills in places which had special things to offer. The self-fertilising tidal marshes of the bay of Fundy were put to good use, first by the French and later by farmers of Yorkshire stock. They brought techniques from home to improve productivity—producing hay to feed the ever-increasing population of horses which both pulled the ploughs and made transportation possible. Gradually, an ever-broadening area of land was tamed and brought into agricultural production, but not without problems both from indigenous and introduced species. The American Dream contained its fair share of the elements of nightmare.

Exactly when the first boatload of explorers from the Old World clapped eyes or set foot on the New World is open to much transatlantic debate. Egyptians on rafts around 5,000 years ago? Lost fishermen from Japan? Celts in leather boats? Leif Ericson and his Vikings a thousand years ago? Portuguese explorers in the 1200's? Or Christopher Columbus in 1492? All have their ardent supporters. However, none of them can claim either the act or the fact of discovery, for long before they arrived the whole of the broad estates of both the Americas had been explored. By the time Columbus landed, many of its sustainable natural resources had been put to use, all its soil types had been tilled and most of its potential crop plants, both cash and food, had been developed to a high degree. What is more, civilisations, each with their own laws, rituals, arts, crafts and even sciences, had come and gone, leaving their own distinctive marks upon the countryside and on the customs of its people.

All this had happened within the New World with no outside help, no Gods in Chariots, no mighty invasions of armies or cultures. The real discoverers of America did it on foot across Beringia and, once the land route finally closed behind them, they were left in peace to get on with the job of colonisation.

Whether the chance encounters by the mariners mentioned above had any real effect we shall probably never know. At the very worst, such brief encounters could have brought diseases like influenza, common cold, cholera and smallpox to a population which had no natural resistance. Epidemics would have been most catastrophic in close-knit 'urban' societies and could have accounted for the rapid demise of certain of the cultures. The survivors could have been those small groups still living a more nomadic isolated existence, but this is only a surmise.

One thing we know for certain is that such waif contact came to an abrupt end when on the 21 December, 1620, European settlers in the guise of the Pilgrim Fathers landed on Cape Cod and, calling it New England, founded Plymouth, which became a beach-head for their conquest of the continent.

They were in reality, refugees from parts of the Old World, which was so over-flowing with human beings that new horizons and, especially, new resources—game, timber and, above all, good soils—had to be found.

The New World, with its 'primitive' people, who still hunted with nothing more sophisticated than stone-tipped arrows and spears and knew nothing of the use of iron, was there for the taking or making. Those two simple words span the whole range of attitudes of the settlers who poured in from then on: the conquering element, who did not mind if they pushed the 'savages' to extinction and took over their lands, and those who genuinely believed that they could improve the lot of both themselves and the locals by showing them the light of European culture and religion.

History has documented and still is documenting the story, and as a 'taker' or a 'maker' you must judge for yourself the outcome to date.

I, however, suggest that, before you do, you go back in time and see it all for yourselves and, thanks to the American flair for showmanship, it is possible to do just that.

Pilgrim Plymouth and the adjacent Wampanoag Indian summer encampment have been reconstructed in painstakingly authentic detail almost on their original locations. Once you have entered through the stockade gate you step back into 1627, not just in terms of the rough paths, wooden buildings and domestic plants and animals, but also in the terms of the people you meet. 'They portray through dress, speech, manner and attitudes known residents of the colony in 1627. Their lives follow the seasonal cycle of all farming communities—planting and harvesting crops, tending animals, preparing meals and preserving food. Busy as they are, the villagers are always eager for conversation. Feel free to ask questions, and remember, the answers you receive will reflect each individual's seventeenth-century identity.'

The above is a quote from the official brochure—and it really does happen, a unique and rather strange experience, for if you ask a question which does not relate to their time scale, they just turn away.

Reconstruction of a house and garden at Pilgrim Plymouth.

Likewise, a walk through the Wampanoag encampment takes you back even further into that period of pre-history when the modern Indian tribes lived almost as part of the natural system.

'On the banks of the Eel River, on the centuries-old site of a similar encampment, an extended family—parents, grandparents, aunts, uncles and children—have built their summer dwellings. The adult men and teenage boys will hunt, fish and tend their tobacco fields. The women and grandparents will care for the younger children, constantly teaching while they cultivate crops, weave baskets and bags for storage, preserve food and prepare the daily meals.

'The seasonal routine is marked by special events. Members of neighbouring Wampanoag families are entertained with food and games. A few colonists arrive to trade for food or to discuss.' (Sic)

Another quote from the official guide. But there is nothing official about the scene you walk through; it is a real experience about real people.

The fact that the settlement and encampment are adjacent to each other and that there is amicable contact between the two groups shows that each had something to offer the other. A look in the neat gardens behind each settlement house confirms this, for there are Corn, Beans and Squash growing alongside more European plants.

It must be remembered that the settlers had left a Britain which was divided into regions, each with its own trends, traditions and developments, even when it came to types of gardens. Around such major cities as Norwich, Canterbury Colchester and London, which were influenced by the Continent of Europe, gardening had developed to a much higher degree than in the more rural areas of the north and west, where it was still in a state of almost medieval simplicity.

Advanced gardens belonging to the Aldens, Brewsters, Cooks, Fullers, Warrens and Winslows consisted of separate plots, one for each type of plant. In one, sweet-smelling herbs to be used in cooking and for strewing on floors and laying between freshly-washed linen were growing; in another, physic herbs for medicine; in another, pot herbs for sallets and cooking; yet another with sallet herbs and roots, which we would call vegetables. All in order, all well tended, each growing in a raised bed edged with rough timber. Cultivation of the more basic gardens belonging to the Allertons, Bradfords, Billingtons, Hopkins, Howlands, Soules, and Standishes is much the same, but the beds are less well ordered and the whole effect is more 'down market'.

Without doubt, the strangest thing about both types of garden is the method of application of fertiliser. Muck or compost heaps, which contain both kitchen and animal waste, are an important part of each garden and help in the constant recycling of nutrients from year to year. However, at certain times of the year when fish were plentiful (and in the early days they certainly were), Alewives and/or Herring were used, one being planted alongside each plant and left to do its rotten best, an ideal, aromatic, 100% organic, slow-release fertiliser. This was a trick learned from the Indians, who also cultivated raised beds—and it was especially necessary when the all-demanding but all-sustaining Corn was planted. Corn rapidly became an important crop for the settlers, along with Beans and Squash, the latter in the form of Cucumbers and Melons, a blending of the know-how of two cultures.

It is of great interest to go and see it all for yourselves and to compare what you see in the gardens with the plants you would find growing in an English country garden today.

PLANTS GROWING IN THE BRADFORD HOUSE GARDEN, PLYMOUTH, NEW ENGLAND

Barberry, Common	*Germander*	*Pot Marigold*
Basil, Sweet	*Heartsease*	*Rue*
Borage	*Lavender Cotton, Grey*	*Sage*
Bouncing Bet	*Lavender Cotton, Green*	*Sassafras*
Catnip	*Marjoram, Sweet Knotted*	*Savory, Summer*
Columbine	*Mints, Peppermint*	*Savory, Winter*
Costmary	*Spearmint*	*Southernwood*
Elecampane	*Mullein*	*Tansy*
Flax	*Mustard, White*	*Thyme*
		Wormwood

All except Sassafras were brought to the settlement from the Old World.

It is also of great interest to see how the settlers and their descendants developed the use of certain local plants. One such plant grows in its wild state on the acid peatlands which are a feature of the eastern coast, clear down from the Tundras of northern Labrador to south of New York: its name, *Vaccinium macrocarpon*, Cranberry to the white settlers, each Indian tribe having its own name for its delicious and useful berries.

It is believed that Cranberries were used at the first Thanksgiving Feast in 1621 and, if they were not, they were certainly present in all their bitter-sweetness growing on the acid soils nearby.

The Indians used them in many different ways: as the source of a dye for their rugs and blankets; a treatment for wounds inflicted by poison arrows; and as a constituent of one of their favourite dishes, pemmican. In its original form pemmican was a mixture of dried venison, fat and cranberries, patted into cakes and dried in the sun. It has since been copied in many different forms for, apart from being very nutritious, it has the added advantage of storing and travelling well, and is still a mainstay, along with Kendal Mint Cake, of certain modern expeditions.

Wild cranberries soon became popular with the settlers and each fall whole families would go out to pick sufficient to preserve for winter. So popular did they become that in 1773 one community on Cape Cod levied a dollar fine for anyone picking more than a quart of cranberries before 20 September. They also became an article of transatlantic trade, for to appease King Charles the Second's wrath for their coining the Pine Tree Shillings, the settlers sent him, among other local produce, ten barrels of cranberries.

During the days of the clippers, when sea voyages could be very long-drawn-out affairs, ships of the American lines often carried barrels of cranberries to help ward off scurvy—a condition now known to be brought about by lack of vitamin C in the diet. Perhaps this is one reason why, when cranberry culture became big business, retired sea captains living on Cape Cod took up the profitable business of running a bog! They financed their cranberry ventures in much the same way in which they had financed their ships, by selling one sixty-fourth shares in the project.

It was one, Henry Hall of Dennis, Massachusetts, who in 1816 first realised that cranberries grew both bigger and better on coastal bogs where they were subject to the gentle rain of blown sand. This observation became the firm foundation for the cranberry industry, which has been a growing concern ever since.

All that it takes to grow cranberries is acid peat, fresh water, sand, the right climate and the initial investment. An area of acid peatland is cleared of its natural vegetation and is well drained before being levelled and covered with a good layer of sand. The cranberry vines are then planted and irrigated, after which they are left for the local environment to do its best. It takes from three to five years for a new bog to come into production and then, with regular weeding in the spring, pruning in the fall, and fertilising and resanding at intervals, it will go on producing for decades, if not centuries.

The first commercial cranberry bogs were in production on Cape Cod in the

1820s and there are still over 400 growers active in the region. From there the new crop spread with great rapidity to New Jersey in 1825, Wisconsin by 1850—where the 1856 crop at Berlin was so good that it couldn't all be gathered in—Washington on the Pacific Coast by 1883 and then on to Oregon and British Columbia.

During this expansive time the growers were always on the lookout for the biggest and the best producers and especially for varieties which were specially suited to local conditions. This process of unnatural selection gradually produced the hundred plus varieties now known to growers. Some of them have wonderful names, like Potter's Favourite, Budds' Blues, Centennial, and Aviator, recording the achievements both of the industry and the times.

Despite all this variety, today four main types are used: Early Blacks with dark red berries which, as their name suggests, ripen early in the season; Searle's, because they were first grown by Andrew Searle in 1894, ripen to a deep red, but later; McFarlin's, named after the man who first took the commercial strains to the West Coast, are deep red, almost oblong in shape and ripen mid-season; Hayes' are the standard late-season variety, producing a medium-red berry which is larger (and, I reckon, tastier, especially when it has been lightly frosted) than the Early Blacks. As recently as 1940 a new hybrid was found at Whitesbog, New Jersey, a cross between a McFarlin's and a Potter. It is a mid-season plant and an excellent yielder, which is now being extensively planted in all growing areas.

Yes, this is an industry which, though firmly rooted in past tradition, still has a bright and growing future with a projected production for 1985 of over five million barrels; whether that will be sufficient to meet the increasing world demand is anyone's guess.

Apart from the actual varieties grown, the methods of harvesting have probably been subject to the greatest change. In the early days, a large wooden scoop was used, with a leading edge like a gigantic broad-toothed comb. These were pushed through the vines by the harvesters, who crawled along behind gently rocking the scoop to dislodge the berries—one reason why an important factor used in selecting the best crop plants was the time the fruit took to ripen and the ease with which it could be shaken from the vines.

As mechanical harvesters gradually took the place of the scoops some of the camaraderie of the families of scoopers disappeared. Today the most fascinating and colourful method to watch is the wet harvest method. As soon as the fruit is ripe and ready, the bogs are flooded with sweet water to a depth which just submerges the vines. Machines called water reels or egg beaters, which look like something straight from a Rowland Emmett cartoon, are then steered across the crop. The churning of the paddles agitates the water just enough to dislodge the berries, which then float to the surface, a gorgeous crimson mass which, when stirred up into ordered rows, looks just like the main part of the Union flag. The floating mass of berries is then corralled with floating booms towards the pickup area, where they are scooped or sucked up from the surface. They are then graded, packed and frozen for world-wide distribution. Others are turned into a variety of mouthwatering products, the list of which, thanks to the forward-looking Ocean Spray Co-operative, increases year by year.

Mechanical as it all now is, with automatic sprinkler systems to evade frost damage, long booms called Brooklyn Bridges which move back and forth on tracks for fertiliser application, pest control and pruning operations, it still has much of the magic of a rural industry. Important in the economy of each bog are the local insect-eating birds which help in pest control and the bees which pollinate the pale pink flowers that open in May in such profusion that they appear to dust the vines with pink powder. Though minute, each flower looks like the head of a Crane and that is how this plant, spirit of peace and thanksgiving, came to get its American name.

It was not only special plants which became important to the settlers, but areas with special characteristics were settled first, especially if these characteristics were similar to those back in the homeland. One such area was the Bay of Fundy in New Brunswick, just over the border of what is now Canada and north of the Gulf of St. Lawrence, which marked the northern boundary of corn cultivation by Indians.

The Bay of Fundy boasts one of the highest tidal ranges in the world, a pulsating 11.58 metres at neaps and 13.87 metres at springs, with a record in 1869 of around 21.34 metres. It is impressive to stand at Moncton and see the tidal bore rush up river, bringing with it large blocks of ice to be deposited along with reddish mud on the marshes. At the head of both Shepody Bay and Cumberland Basin, which themselves form the head of the Bay of Fundy, no less than 30,000 hectares are covered with a wonderful complex of tidal march, freshwater swamp and sphagnum bogs, a real Bellamy paradise!

'From a high vantage point upon the smooth, rolling hills which surround the basin, the marshlands can be seen in all their glory. They are seemingly as level as the sea and, like the sea, they fill bays, surround islands and are pierced by peninsulas. They are treeless, but are clothed nearly everywhere with dense swards of grasses in many shades of green and brown, the colour varying with the season, with the light and even with the wind. Frequent ditches marked by denser growth, rare fences and occasional roads or railways are the only signs of the operations of man. Towards the sea are narrow fringes of un-reclaimed marsh, poorer in vegetation and generally duller in colour, while further back the marshes give place to the browns and greens of the bogs, which are further distinguished by irregular shrubberies of stunted trees and many little lakes.

'Nobody lives on the marshes, but scattered upon them are many great [and I mean great!] barns all of one and the simplest of patterns, unpainted they stand, grey with the weather, at any and every angle. These barns are one of the distinguishing features of the marshes and give to them a suggestion of plenty, which is a true index of the economic condition of this region, for here are the most progressive farmers and the most thriving country towns in Eastern Canada.'

The above description is adapted from an account of the marshes by the famous plant ecologist W. F. Ganong, published in 1903. Today the scene is much as he described it, but many barns have gone and the economy is far from good.

Close inspection of the marshes (and it can be a lovely muddy experience!) reveals that, apart from being very extensive, they are of a special type, composed entirely of inorganic mud, in places 25 metres thick. The source of the mud is the soft red sandstone which forms the sides and bottom of the channels which bring in the water from Fundy Bay—and the motivating power is the rush and scour of those record tides. As the seaward channel is scoured, so the level of the marshes has been built by deposition, a process which could not have continued for long but for the fact that the land itself is gradually subsiding.

The great weight of the Wisconsin Ice Sheet so depressed and warped the earth's crust that it is still heaving a sigh of relief and rising to take up its original position. As the inland areas go up, the coastal sections, which were not weighed down by ice, subside and will continue so to do until a new equilibrium is reached. Along this section of the coast, as the coast subsides, each high tide, and there are two each day, brings a new layer of mud and, like a master plasterer, spreads it across the marshes, thus keeping their head almost above water.

Fresh water flowing down into this ever-full arm of the sea, ponds along the landward edge of the mud, producing ideal conditions for the development of freshwater swamps and bogs. Proof positive of this continual subsidence and build-up is found in extensive sub-fossil forests of Beech and Pine, and peat layers, beneath the mud. They date to between 1000 to 2000 B.C., when subsidence must have become effective in relation to the post-glacial rise of sea level.

The natural vegetation of the marshes at the time of first European settlement in 1670 consisted of sparse, salt-tolerant communities of Cord Grass, Samphire, Sueda and Statice. The first settlers were French and they came from Saintonge on the west coast of France, where they had farmed tidal marshes. They called their new homeland Acadia and farmed its marshes in their traditional way.

To reclaim a marsh for agriculture, three things must be done. First, the sea must be shut out. This was accomplished by means of dikes, constructed of mud with a core of stakes and brushwood and protected from tidal scour by lines of stakes, piling or loose stones. Secondly, the salt must be removed from the mud by allowing the rain to drain through it. Thirdly, drainage of the developing soil must be established, not only to speed the removal of salt but also to separate it, thus allowing a soil fauna and hence a good structure to develop. The latter was brought about by ditching, each ditch being no more than half a metre deep at its landward end, but deepening all the way to flow out beneath the dike. The seaward end of each ditch was terminated by a wooden sluice, across the mouth of which hung a wooden 'clapper', hinged at the top and inclining outwards, that is towards the river, at the bottom. During periods of high tide, the clapper door is pushed shut, thus excluding the mud-laden salt water from the dikes and hence from the marsh. Fresh water is at the same time ponded back in the dikes, but not for long because the mouth of the sluice is set just below high water mark. When the tide goes out, the clapper swings open due to the press of fresh water and the soils are drained of water and salt.

When a sluice of this sort is used in a dike thrown across an entire river, so that sea and mud are excluded from the whole river system all at one go, the structure, dike and sluice, is then called an *aboideau*. The method was very effective, as it had been back home in France; and a stable soil and a good crop of grass soon became the order of the Bay.

Once reclaimed from the sea, the marshes were very fertile and Ganong claimed that they were unsurpassed, if equalled, by any other land in Eastern Canada. The best marshes may be cropped with undiminished yields for many decades without the need for fertilisation, and it is reported that some of the areas have been continuously cropped in this way for over 200 years. One further feature of great importance is that, in a dry year, yield from the marshes is not as badly affected as that from the upland meadows which soon loose their available water.

The area around Sackville thus became a centre of Acadian settlement for two reasons. Firstly, because it contained the largest expanse of reclaimable coastal marsh and, secondly, because of its situation at the Chignecto Isthmus which, except for Portage Hill, offered a navigable connection between the Bay of Fundy and the Gulf of St. Lawrence, a cheap trade route. The marshes could provide much more than the locals and their animals could consume, while horse-drawn transport in the expanding villages and towns of the colony could consume more than the local soils could produce and, in the absence of good roads, waterways were the best and cheapest method for speeding the hay and other marshland products on their way. Also, New Brunswick became a centre of boat and ship building.

The local Acadians were expelled on the eve of the Seven Years' War in 1755, and the land was re-occupied in the following decades by settlers from New England, which was rapidly filling up, and from Yorkshire in Old England. Many of the descendants of these people have remained there ever since, farming the marshes.

Early in the nineteenth century, a native of Sackville, of Yorkshire descent, by the name of Toler Thompson introduced a technique from back home, which allowed the extension of reclamation back into the area of freshwater swamps which fringed the tidal marshes. The technique had been developed in the English Fenlands and especially those south of the Humber on the borders of Yorkshire and Lincolnshire.

Canals were dug into the swamps, through which salt water and its muddy load were pushed up by the tide. These areas were thus turned, temporarily, into tidal marsh, killing all the freshwater vegetation and adding nutrient-rich mud to the organic mulch. When enough mud had accumulated, reclamation was concluded by the usual diking and draining. Likewise, controlled access of mud-laden water was used to rejuvenate the soils of old agricultural marshes which had passed beyond their prime. This process was called 'tiding'. It was during this time that the construction of *aboideaux* controlling whole river systems came into great question and even local conflict, for once constructed, 'tiding' of small sections became a much more difficult process.

Vegetable, and some root, crops were grown on the marshes, but did not do well enough for real commercial development. Likewise, there was some production of cereals, mainly Oats, but emphasis became more and more on management solely for the production of hay. The agricultural marshes behind the dikes included two kinds of hay meadow. The 'broadleaf' meadows dominated by the native Cord Grass were held in high esteem and their products kept for local use. The 'tame' or English meadows were dominated by Timothy Grass, Couch Grass, and various admixtures of Red Clover, all introduced from England; they grew on the better-drained sections and produced hay mainly for export. After harvest it was stored in the large barns until winter, when it could be transported by sled to the ports and in later times to railway freight yards.

It was the first two decades of this century that were the real 'hay days' of the marshlands when, to keep the million-horse towns in running order, hay was exported to Boston, New York, Venezuela and even South Africa. I wonder what happened to all the horse exhaust that flowed onto the streets in those days? Perhaps the emissions from the horseless carriages which replaced their grace and elegance are no worse a hazard!

The development of 'on-line' production by Henry T. Ford brought to an end this period of expansive effluence and local affluence for the farmers who managed the marshlands so well. During their hay days, the marshes served, not only as energisers and fertilisers for far-off places, but also for the local uplands. Hay grown on the marshes was fed to cattle who manured the upland pastures. In addition, marsh mud, which was delivered free and in profusion twice each day, was carted up and spread upon the fields.

It is still possible to get the feel of those days of affluence by staying at the Marshlands Hotel in Sackville, where you will sleep in 'four posters' and enjoy a taste of the elegance of life at the turn of the century. You can also walk the windswept marshes, witness the pulse of that life-giving tide, red-brown with mud, in diastole it overflows the banks, slows, hesitates, stops and, having shed its load returns, not crystal but almost clear, the way it came. You can also see the great barns and covered bridges, or at least what still remain, now much greyer than when Ganong described them almost 80 years ago, and sadly falling into decay. But please go quickly, for time and tide wait for no man!

Every part of this great land mass of North America has its own special characteristics which are, at least in part, mirrored by its natural plant and animal communities. It must have been very exciting to have been a pioneer settler moving out into virgin territory to set up a new home. What rewards and what problems lay in store!

Only ten years after the Pilgrim Fathers had landed on Cape Cod there was a bounty on the head of the Grey Wolf, not because it was a hunter of man—it never was—but because it attacked farm animals and dug up the fertilising fish planted in the corn patch during the fourteen days before they rotted. A palisade was built around the settlement and fences seven feet high were erected around fields to keep Deer at bay. The predators of the forest soon responded to this new oasis of goodies and all became specialists in their own field; Bears for hogs, Cougars and Panthers for cattle, and Foxes and Wildcats for poultry.

The Indians protected their fields by constant watch from raised platforms set amongst the crops. Birds, especially Crows, Redwings, Blackbirds and Grackles, were particularly troublesome at sowing times, and were the main reasons why a very large sowing-to-yield ratio had to be maintained.

The white settlers did not move into a primeval landscape; much of it had already been hunted and even farmed. There were, for example, hundreds of Indian villages in New England, some of them with up to 70 hectares cleared for crops, which were planted around the intractable tree stumps. The surrounding woodland was managed for hunting by burning the undergrowth, a practice which provided fresh young growth to tempt the game—and gave the hunter a much clearer view. The village sites were, of necessity, moved every ten to fifteen years, the reason being as much the distance the squaws had to travel in order to find firewood as any loss in soil fertility, for it was they who developed the fish fertiliser technique. As they moved on, the forest moved back to retake its place, although the resultant secondary forest was often rich in Pine and very different from that which was originally on the site.

Just how much unaltered forest was left, even in 1621, is impossible to say. Large areas of secondary forest were already in existence where slash-and-burn cultivation had passed that way. Likewise, the make-up of both the canopy and the ground flora of the fire-managed game areas was much altered, such management favouring fire-resistant trees and shrubs, perennial herbs with underground storage organs (rhizomes, bulbs and corms) and even annuals. The unburned forest round about would have been subject to specific collection of nuts, fruit, storage organs and other parts of plants for use as food and in crafts, medicine and ceremonies.

It was observed in 1635 that the managed forest was different; silent by day, its birds were Ravens, Owls, Eagles, Passenger Pigeons and Carolina Parakeets. With increasing deforestation all these became less common and some extinct. On the other hand, many of the song-birds which are, even now, so common, cannot live under the conditions of mature forest and so they increased with deforestation. They were, however, especially abundant during the earlier phases of settlement, when small fields were spreading into the forest, creating more and more edge, scrubby vegetation which provides them with an ideal habitat within easy reach of food and shelter. It was in this 'edge' or ectone (a border area between two types of vegetation) that many different weed species found a permanent home, from which they could move in to invade the cropland. Many of these weeds were brought from the Old World: plants like Plantain (known to the Indian as Englishman's Foot), Corncockle, Saint John's Wort, Couch Grass, and Charlock (called locally Terrify) and many more.

It is of interest to note that the transformation of forests into fields at similar latitudes in Europe had taken more than 2,000 years, ample time for the development through selection of such opportunistic plant species; also animals, for with the settlers came House Mice, and both Brown and Black Rats. No wonder that in America, when the same process took place in around one-tenth of the time, it was the weeds from Europe that won the day.

The settlers also soon found that the grasses of their new North American woodland home were neither trample-tolerant, nor were they of such high nutritive value for feeding livestock as those they had depended on back home. Free-range cattle were forced to browse twigs from trees and often starved to death in winter, and sheep did little better. So grasses and clovers were imported and the pastures were soon much improved. This meant that more cattle could be fed and more draught horses could be kept, and so the destruction of the woodland and the planting of new crops went on apace. It did not take the wildlife in each new locale long to realise what was happening and cash in on the fact, for as the woodlands became more and more dissected by fields, the edge effect increased and with it came more and more problems.

It is a basic law of ecology that there is always a net flow of energy and raw materials from the less to the more mature system. Farming, in essence, taps into young pioneer ecosystems created by man to give a fast return; when these are surrounded by the maturity of climax, or even secondary, woodland, with all its abundant and varied forms of animal life, the rule must be obeyed. The forest animals seize the opportunity of lush food with thankful beaks and jaws.

So bad was the situation that there seemed no way of stopping the depredations of Grey Squirrel and Raccoon, they just came and kept on coming, especially when the nut and fruit crop failed in the woodland round about, as it did at intervals. In northern Pennsylvania this seemed to happen with catastrophic regularity every seven to eight years and, when it did, the Squirrels marched in their thousands to the south-east, laying waste the grainfields. With bounty at three pence per head in 1749, the state spent £8,000, exhausting the treasury, and the farmers grumbled that their workers were leaving them to hunt the varmints. In 1784 a special tax had to be levied to raise the necessary squirrel-bounty money, and similar problems can be quoted from Maryland, Virginia and many other states as they were settled. To complicate the farmers' lot, many of the weed species went ahead of settlement, finding their own niches on riverside gravels, cliffs and abandoned Indian Farms; and others followed on behind. During early settlement days in Michigan there were no Crows to steal the Corn before it rooted, no House Mice to eat the stored grain and no House Flies to harass the farm workers; all these arrived in force once sufficient woodland had been destroyed.

It was no simple problem. In New England a bounty was placed on the Grackle, a ground-feeding bird which stole the seed. This was so effective that in 1749 the worms ate up all the grass, so Grackles, which eat both grain and worms, were granted a reprieve, for hay had to be imported into the state that year.

The most dramatic cause for concern were the plagues of Passenger Pigeons, millions of which came down like locusts and devoured everything in sight. In 1642 they beat down the grain around villages in Massachusetts, and well into the nineteenth century they sometimes consumed so much of the woodlands' natural produce, beechmast, acorns and chestnuts, that the settlers' free-range hogs died of starvation. There are, indeed, stories of so many of these birds attempting to perch on trees that branches snapped off, and even mature trees toppled over. Whether the tales are just 'pigeon pie in the sky' I do not know, but I do know that by 1923 the menace was past; the species was extinct. It could not have been simply pest control measures which brought about their final collapse, for at the best these killed off the weakest and dispersed the others to breed in the most advantageous habitats. The most logical explanation is that their numbers increased with settlement and then crashed as their most advantageous mixture of habitats, woodland adjacent to well-kept fields, declined.

It is estimated that when the first European settlers arrived, the sum total of woodland, from the Atlantic to the Mississippi and from the 47th parallel to the coastal plains of the Carolinas, was in excess of 174 million hectares; today 7.7 million hectares are left. South of this main temperate forest belt the imported grasses and clovers did not grow so well and further south, in the Virginias, the lack of a true winter rest period made their growth impossible. Hence, south of the Mason-Dixon, or clover, line, it was found very difficult to fatten the enormous number of cattle, there just was not enough hay for winter feed. By the early 1700s the pastures in Pennsylvania and New Jersey were overstocked and deteriorating, and irrigation had to be employed in order to keep up the supply of winter feed for the cattle and horses. So it was that as the population increased and new settlers arrived, new country had to be opened up. It was during such times that areas like the Bay of Fundy, with special characteristics to offer and new crop plants to develop, became of more and more importance.

It must be remembered that, although this was no primeval landscape, there were no roads and, what was most important, no maps. Thus it was that the explorers and trappers went ahead into the unknown, and the settlers followed, with always the hope of an agricultural Eldorado at the end of the trail.

Two natural products spurred the pioneers on, Beavers and Ginseng, and both were products of the woodlands. The former lived in the woodlands, feasting on the young shoots of Willow and Aspen and felling small trees to build their dams and raise their families in the safety of moated lodges. Before man arrived on the scene, apart from natural fire they were probably the greatest enemy of the woodland, not because of the amount of wood they consumed and cut, but because of the flooding they caused. A Beaver dam placed in a strategic position could back up sufficient water to flood large sections of good creek forest. This had its advantages, creating niches for species typical of more open land, and acting as focal points for game to drink and to take shelter in during forest fires. Such Beaver ponds also acted as firebreaks and so must have been of great importance, especially in the drier West.

Beaver pelts are impervious to water and hence were soon in great demand for the manufacture of hats and from the early 1600s through to the middle of the 19th century no proper European gentleman would be seen out and about without a beaver hat either on his head or in his hand. For the first 150 years

The presence of Beavers spurred the pioneers to open up the New World.

the Taigas of Canada kept this world of hat couture supplied and the majority of the exploration of the famous Hudson Bay and North West Companies was floated on the profits of this aquatic mammal. However, with fashions catching on in the New World and a beaver hat fetching $10 in New York, fortunes could be made much nearer home. The trappers thus became trail blazers on the muddy Missouri River and beyond, fanning out from the port which soon became the city of St. Louis.

Ginseng is one of those woodland plants which, above all, characterises the Eastern Connection. It is a perennial, about 15 to 20 cms. tall, with 2 to 4 leaves that are divided into five leaflets which are arranged like the palm of a hand on the end of each stalk. The small terminal stalk produces inconspicuous flowers and red berries. Its roots are large and aromatic, and are sometimes shaped like the body of a human; and this is probably how its legend arose, for it is regarded in the East as a cure-all. The roots are also used as an ingredient of love potions and talismans. The fact that its therapeutic or aphrodisiac properties have never been proven in the West does not matter; hundreds of generations of Orientals cannot be wrong and, as American Ginseng was, in their eyes, as good as, and some even consider better than, that grown at home, it soon became a lucrative trade; as it is still, being exported to Hong Kong, Singapore and many other places across the world. So it was that the trappers would return down the tributaries of the Mississippi, back to their base at St. Louis with packs laden with Beaver pelts and Ginseng roots.

It was as they gradually edged west, reaping rich rewards from both their choice commodities, that they must have heard tell of a boundary somewhere up ahead where woodland gave way to wide-open fields: the Prairies. This was a very important boundary for the trappers, for beyond the confines of the woodland there were no Beavers and no Ginseng. It was also to be a boundary of immense importance for the settlers who followed on. The first published map which depicted boundaries of the major types of vegetation in America accompanied a paper by Pickering, 1830, entitled 'On the Distribution of Plants'. There were, however, even then maps in existence which depicted the position of the all-important 'wood edge'. To quote from a paper by G. Malcolm Lewis, 'A manuscript map presented by Alexander the Elder to Sir Guy Carlton, somewhere between 1776 and 1778, shows a boundary extending for approximately 1,000 miles from the Red River around 49° North towards the Rocky Mountains in approximately 53° North. This line was coloured blue or green and labelled "The Course of the Great Plains". To the south and west of this line bold captions indicate "Great South Plains" and "Great North Plains" in what are now respectively North Dakota and Saskatchewan/Alberta. Three of the four surviving contemporary copies of a map first drawn by a Canadian fur trader, Peter Pond, show a similar boundary extending for approximately three thousand miles from a point a little west of the Mississippi-Ohio confluence to somewhere in the Peace River Valley. On one version it is labelled "Ye Eastern Boundares of those immense Plains which reaches to the Great Mountains".'

Knowledge such as this was passed back down the line of contact to the burgeoning populations of settlers waiting their chance for fame and fortune.

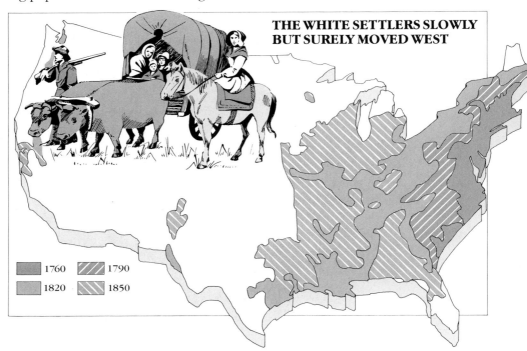

THE WHITE SETTLERS SLOWLY BUT SURELY MOVED WEST

	1760		1790
	1820		1850

CHAPTER ELEVEN

Diamonds and the World's Best Friend

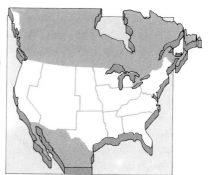

THE effects of the ice age left a broad belt of new young soils and many old ones rejuvenated by change, the potential of which was there for the taking by anyone who had the will and the horse power. All this attracted the new immigrants who were fleeing from an overcrowded Old World ever westwards towards the promise of easy pickings on the broad flat prairie lands, with no trees to fell. Problems like introduced weeds and disease went before them and in time were solved by the increasing use of science, machinery and fossil fuel. The markets ever widened and more land was opened up to answer the demand for flour power. Unfortunately the cost of this American Agricultural Machine is now so great in terms of fossil fuel that far-reaching changes are on the way.

In 1863, Peter Davis of Washington Township in Indiana found a flawless diamond in Goose Creek.

In 1865, John Ward started to drill for oil on Manitoulin Island in Lake Huron.

In 1918, a farmer of Crystal Bay, Minnesota, destroyed a hedge of 635 Barberry bushes.

In 1982, the world looks to the wheat belts of America for its daily bread.

These seemingly unrelated facts constitute the key to an understanding of an American revolution which is still going on, a revolution on which the future of the world has come to depend.

Diamonds are made of carbon—and so are all living things. In the former it exists in its pure elemental state, while in the latter—and remember that includes you and me—the carbon forms the connecting link between a whole complex of compounds which include oxygen, hydrogen, nitrogen, phosphorus, potassium and certain other minerals. Diamonds are only formed where the conditions are exactly right, that is where there is immense heat, enough to melt the element, and immense pressure during its subsequent cooling and crystallisation. It is the hardest substance known on earth and so can be used to cut through any other part of the earth's crust. Living carbon can only exist where the conditions are exactly right, that is where water and minerals occur together in the right environment, along with a genetic message which has been programmed through evolution to make use of that environment.

THE DIAMOND NECKLACE
The ice sheets of the last great glaciation left America with a diamond necklace
and the world with much more – hope for a well-fed future

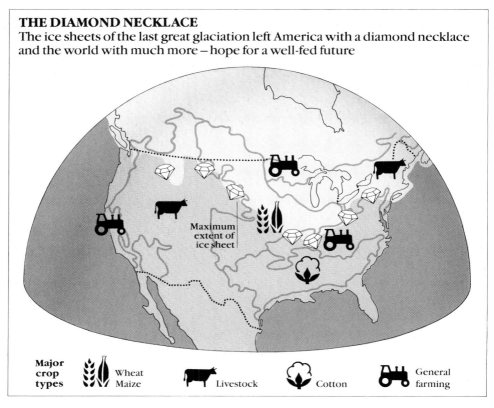

Major crop types — Wheat / Maize — Livestock — Cotton — General farming

The genetic message which is *Homo sapiens*—you, me and the four billion other human beings on earth—has the capability and dexterity to manipulate many of the earth's environments to provide for both its needs and its excesses. In the hands of man, commercial diamonds have opened up an immense store of fossil energy which can, like the best cut gemstones, be used for good or evil.

Since Peter Davis' discovery, at least 48 other similar diamonds have been found in America and Canada, and the tale of each one begins with jubilation and often ends with intrigue and eventual theft. How many more have been unearthed and lie unrecognised for what they are is anyone's guess, for the raw stones look nothing like the finished product cut by a skilled craftsperson.

The known localities of these drift diamonds stretch in a ragged line, south of the Great Lakes, across Indiana, Wisconsin, Michigan, Illinois, Ohio to Ontario and New York. For a long time their origin was a source of great speculation, for among the many bountiful resources of America, large deposits of diamonds are unknown. It is now thought that they had their origin in the region of James Bay in Canada, from whence they were transported by the various lobes of the ice sheet of the last great glaciation to await eventual discovery. The ice age, however, gave the world untold riches far in excess of those 49 assorted diamonds. It gave the world the bountiful face of America as we now know it.

The world's largest island surrounded by fresh water sits astride the northern end of Lake Huron, separating it from Georgian Bay which boasts at least 10,000 islands of its own. It is a fabulous spot for a holiday away from it all, a botanist's paradise and a perfect place in which to get the feel of the many effects of the last ice age. Local legend has it that the Great Spirit Manitou used ice knives from the north to carve the land into many pieces which he scattered across the great waters. He then gathered the choicest bits to form Manitoulin Island, God's Home. It is thus a place in which legend, pre-history and scientific fact are one, for not only could the ancestors of the Ojibway people have witnessed the melting of the ice, but the land which they colonised bears all the marks of its passing.

The hardest rocks, sandstones and quartzites, mark the southern boundary of the Pre-cambrian or, as it was known in my schoolboy atlas days, the Laurentian Shield, which itself formed the heart of the ancient super-continent of Laurasia. They have been scraped clean, smoothed and rounded. In places their flanks are polished; in others they are scratched and gouged, the alignments of the marks indicating the direction of ice movement. They form knolls and hillocks separating many small lakes, a topography which epitomises the shield country and paves the way to Manitoulin which is itself made almost entirely of limestone, covered in places by drift material brought in and dumped by the ice.

Dreamer's Rock is one such knoll which stands at the gateway of Manitoulin. It was there that young Indian braves on reaching puberty came to fast, to sleep and dream, and in those dreams to find a place within their own society: hunter, warrior, tiller of the soil. Those who lived on the Great Island had much to be thankful for, for wherever the glacier had left an admixture of well-ground rocky debris full of mineral promise, good deep soils had formed. These supported a diverse forest of White and Red Pine admixed with Sugar and Red Maple, Eastern White Cedar, Red Oak, Basswood, Trembling Aspen, Yellow and White Birch, Balsam Poplar and many more, a forest rich in game and a profusion of smaller plants including no less than four species of Lady's Slipper Orchid—Stemless, Ram's Head, Showy and Yellow, the latter being the same species which used to be abundant in certain parts of Britain and now restricted to tiny colonies in two separate secret localities. Here it is so common and plentiful that, wait for it, I picked one! It was in the path of a caterpillar tractor and I moved it out of certain destruction into the sun where we could film it in all its beauty. I did this under the watchful eye of John Morton, local expert and Professor of Botany at Waterloo University, who has compiled the flora of this truly floribundant island. To date it includes more than a thousand species of vascular plants, almost one quarter of the total flora of Canada. What is most exciting is that it includes species drawn from all corners of North America, and more are being added to the list each season.

Manitoulin may be the biggest fresh-water island in the world, but whence all this diversity? The whole area is still recovering from the effects of the glaciers, which did their work to perfection, and the vegetation has not yet completely stabilised. Although some parts are covered in deep drift material which has developed good soils and mature forest, in others the rock is still as bare as the day the ice scraped it clean. Most striking of all these open areas are the limestone pavements. Limestones are sedimentary rocks which by their nature were formed in layers. The ice, which was about one mile thick (I am not going to metricate its awesome presence), simply planed it as flat as any pavement. Hence the name and the terrain, which looks more like the runway of a disused airport than any natural phenomenon. One of the weirdest plants to inhabit these flat open spaces is the Horizontal Juniper, which lives down to its name, for with no soil to get its roots into, it creeps across the flat surface, adding its own special magic to the slow process of colonisation. Its twiggy arms reach out across the open pavement collecting as they do so any rock debris, sand, silt or organic matter blown in by the wind, until there is sufficient to form a pad on which lichens and mosses can begin to grow. The pad of skeletal soil so formed supports, in turn, a diversity of pioneer plants including Indian Paint Brush, Blazing Star and Common Pussy Toes. When sufficient soil has collected, White Cedar takes a tenuous roothold, increasing in stature until some extra strong gust of wind topples it to the ground, exposing a flat pad of matted woody roots and a clear area of limestone rock into which they could not penetrate. This is often not the end of the life of the tree, for in most cases the largest branches turn up towards the sky and continue to grow as a multi-stemmed bush which will eventually be lost in the developing forest.

Time is all that it takes to transform even the hardest rock into soil and the barest pavement into forest, but over much of Manitoulin there has been insufficient time for this to happen, for though the ice melted hereabouts 12,000 years ago, much of the lower part of the island has only become dry land within the last 4,000 years. A sheet of ice one mile thick weighs an awful lot—to be inexact, around 16 million tons per hectare. It was sufficient to depress the land surface by as much as 100 metres. When the ice melted, this enormous pressure was released and the land began to rise up. At the same time melted water filled depressions which had been scoured and deepened by the ice, and the Great Lakes began to come into being. At first, glaciers blocked the present outlet to the St. Lawrence, creating a gigantic lake which has been christened Algonquin, and which drained southwards via the Mississippi. Once the ice sheets had melted, lake levels began to fall, and around 11,800 years ago the highest parts of Manitoulin rose above the lake waters as a series of small rocky islets. The shore-lines of this old lake are clearly marked even today on the highest bluffs, and gives some indication of the extent of the rebound of the land surface. The best place to see them is at Cup and Saucer Bluff, the top of which, complete with a wave-cut platform, stands out in bold relief at 275 metres, 159 metres above the current level of Lake Huron. The presence of a similar shore-line, which forms the rim of the saucer at 99 metres above the present lake level, indicates the complexities of the rise of land and fall of water which shaped Manitoulin and the Great Lakes as we now know them. Throughout this time, the size and shape of Manitoulin have changed, and land bridges connecting the island to all points of the compass have opened and closed as lake and land levels reached new short-term equilibria. Standing as it does at a crossroads of migration, the flora of Manitoulin has thus been enriched by immigrants from all areas of the continent, many finding what have so far proved to be permanent homes in the still stabilising plant communities.

It is possible to see many different types, of flowers and even of whole plant communities, growing in juxtaposition: for example, prairie grassland dominated by Old Man's Whiskers with scattered Yellow Whitlow Grass and Burr Oak, a community typical of the drier west. These prairie plants together with many from the warmer south like New Jersey Tea, Blue Cohosh and Beard Tongue, probably arrived on the Manitoulin scene during the warm hypsithermal, and have remained because of the special local environment.

The Great Lakes act as a gigantic storage heater, ameliorating the local climate to such an extent that the first frosts come to Manitoulin as late as 30 September, and the growing season extends almost into November. Add to this the facts that 200 cms. of snow provide winter protection, and that the thin well-drained limestone soil warms up rapidly in the spring, and you have perfect conditions in which these southern plants not only survive, but thrive.

The warm humidifying presence of the Great Lakes must also account for the existence of another group of plants on Manitoulin, best called the Great Lakes Endemics. Pitcher's Thistle, Dwarf Lake Iris and Houghton's and Ohio Goldenrod are species that probably got stranded here after being more widespread during the hypsithermal, while Yellow Flax, Fringed Gentian, Butter Marigold and Hill's Thistle are no more than subspecies, varieties of more widespread species which have become isolated on Manitoulin in the same way.

The northern elements, which include Alpine Poa, Deer Sedge and Swamp Birch, are relics of the time when Manitoulin was close to the edge of the melting ice. They still find a niche in the more open spots, indicating that lack of competition is of equal importance as correct climate when it comes to survival. I would myself include two other species within this northern list: one is the Shrubby Cinquefoil, the other the Bird's-eye Primrose. Both are common on the marsh soils of Manitoulin and both, or at least the former and a very close relative of the latter, occur in Upper Teesdale in northern England, one of Europe's foremost localities for relics of both late and post-glacial conditions.

The final element of Manitoulin's flora is perhaps the strangest of all, for it consists of plants which have their main distribution many hundreds of kilometres to the east, beside the Atlantic Ocean. All have a salty tang to their names: Sea Pea, Seaside Spurge, Sea Rocket, and Dune and Beach Grass. They all arrived on the Manitoulin scene some 11,000 years ago when the melting of the glaciers which had blocked the St. Lawrence River Valley allowed the tide to run up over the still-depressed land surface as far as Lake Ontario, and also up the Ottawa River Valley which then provided the main outlet from the Great Lakes. Bearing in mind the fact that the Ice Age caused the sea level to fall by as much as 100 metres, and that the contemporary sea levels were not restored until some 6,000 years ago, the magnitude of the effects of the Great American Depression may be appreciated in full. Eventually the present friendly face of America, complete with the world's largest waterfall, Niagara, and the world's largest lake, Superior, came into being, but not before many other changes had taken place.

Throughout the glacial period there existed an enormous lake which covered much of the area that now constitutes the Prairie Provinces of Ontario, Manitoba and Saskatchewan, and the States of Minnesota, North Dakota and a small part of South Dakota. We know, because more than 500,000 square kilometres of these flat lands are covered with lake muds and other deposits that were laid down in still, fresh water. This immense area, which has been given the name Glacial Lake Agassiz, was never totally submerged at any one time. During the ice age itself the northern part was covered by ice, the melting of which produced so much water that it eroded the southern outlets to such an extent that much of their waters drained away. It thus contracted in the south as it expanded in the north. Later still, tilting of the land and continued erosion opened up new outlets, mainly in the east, and Lake Agassiz shrank even further. Thus at any one time its open waters probably did not exceed 200,000 square kilometres in extent; but even that is a big lake by anybody's reckoning almost three times the area of lake Superior. What with all this—melt, erosion, tilt and upwarping—all its waters eventually drained away, leaving the flat lands covered with the alluvial deposits which provide the clues both to its existence and its past extent.

Glaciation played a crucial part in the shaping of the American landscape.

Usually the final phases in the history of any d(r)ying lake is linked with the gradual encroachment of marginal swampy vegetation and eventual infilling with peat; the whole going over to some sort of swamp forest dominated by Pine, Birch, Willow, Alder or Aspen, depending on the exact history of the final stages of infill. As peat, of whatever type, is only formed under stagnant conditions in which the water is so poor in dissolved oxygen that the leaf and other organic litter and debris cannot decay, each deposit thus becomes a record of its own genesis and so of the demise of the lake. These living history books, however, do much more than record local events for they also receive into their maw the airborne pollen grains and spores produced by plants growing round about, as well as dust produced by local disturbance. As a result they chronicle the events which take place both within themselves and in their catchment areas.

As the waters of Lake Agassiz were drained away, it did not go through the terminal phases of peaty reclamation en masse. There are, however, many such deposits within the lake's original shore lines, and the lake muds themselves hold their own records of the past. Likewise, such peat-filled basins are common in the wetter parts of North America and are especially abundant in terrain which was beneath, or along the margin of, the ice.

In recent years, many of these post-glacial areas have revealed their inner secrets to scientific research, so that we can begin to piece together the full story of post-glacial change.

The actual southern limit of the glaciers is marked out by a ragged line of ice-worked debris, heaped into terminal moraines or dumped as lateral moraines or eskers along its margins and within its melting mass. Determination of the exact limit of the ice sheet is complicated because there have been no fewer than four major glaciations in the past one million years, the advance and retreat of each having been marked by lesser fluctuations, which included rapid melts and surges forward. Add to this the fact that as soon as ice has turned to water it begins to erode the ice-deposited debris and transport them to new locations. Complicated as it is, the evidence of that all-important southern edge is there to find for anyone who can read the landscape signs: moraines, eskers, erratics, glacial dammed lakes, kettle holes, kaims and a plethora of peri-glacial phenomena.

Why is that southern limit so important? North of its convoluted line the living slate had been wiped clean; all life and the soils which had supported it were obliterated, if not completely erased from the surface of the earth. A mass destruction, but at the same time, or at least a little later, a mass rejuvenation, for once the ice had gone, new parent material full of mineral promise was exposed to an improving climate and the process of new soil formation, ready for any plants, animals or persons that could make use of it.

That was not all. South of the deep-freeze line great changes had also been taking place, and they have been recorded in great detail in the peat layers. The ice-age climate had pushed a wedge of tundra, in places perhaps as much as 300 kilometres wide, south of the ice front. It was a treeless vegetation, dominated by Sedges, Grasses, Dwarf Birch and Willow, Mountain Avens and a variety of plants belonging to the Heather family, including the Bear Berry and the Bog Whortleberry. This open vegetation must have been underlain, at least in part, by permafrost, dead ice and permanently frozen ground, only the surface layers of which melted each spring, to re-freeze again the next winter. Under such conditions, the pressures

Typical Tundra scenes: Cottongrass (*left*) and permafrost patterns.

set up by the constant thaw and re-freeze of the upper active layer stir the soil in time, sorting stones from gravel and silt and aggrading them into neat circles, stripes and irregular polygons, all hallmarks of a landscape on the edge of an ice sheet. Such periglacial phenomena can and do persist long after the permafrost has melted away—and their remains may be clearly seen in Washington D.C., central Indiana, western Wisconsin, northern Minnesota and New England. There is thus surface proof that Tundra conditions did exist, although in the latter two areas it was formed after the period of ice maximum. In other places there is good evidence that no Tundra was present, for forest peats and soils are found right up to what was the foot of the live ice.

This is in marked contrast to the European situation, where the belt of periglacial Tundra was both much broader and more continuous along the ice front. The fact that the ice sheet in the continental climate of North America protruded much further south to reach the potentially warmer latitude of 39°N. in Illinois, compared with 52°N. in Germany, is probably enough to account for the difference. Also, the European Alps were set east-west, blocking the southward flow of cold polar air and so ponding its effects back. Both the Appalachians and the Rockies run north-south.

South of the Tundra belt, however broad and discontinuous it was, the soils had also been subject to very great changes. A belt up to 1,000 kilometres wide, that is down to a line running west from southern Georgia, was covered with Taiga, forest dominated by conifers. In the Appalachians the dominant trees were Spruce, and Banks' Pine, with less Spruce in the south. Westwards to Illinois both Pine and Spruce were present in abundance, while from Kansas through the Middle West Pine was absent.

The pollen in the peatland also indicates that the cone-bearers did not reign in isolation; the broadleaf trees were there, at least in isolated patches. However, despite this and the regional differences, the whole was in fact dominated by needle leaves which are notorious for producing raw acid humus which speeds the leaching of the soils.

As the glacial conditions improved, these zones gradually moved north. The Taiga took up its contemporary position between the south of Hudson Bay and the north of the Great Lakes, and was followed by Oak, Ostrya, Elm, Ash and many more broadleaved trees spreading out from refugia both within and south of the glacial Taiga.

The whole pattern of forest life, including that within the soil, changed from one determined by the year-round sombre shade of an evergreen canopy, to a much more seasonally orientated regime. The softer, broader leaves falling each autumn were in greater variety and, in the main, more tractable to breakdown and rapid remineralisation by the soil flora and fauna. Snow, no longer held aloft by an evergreen canopy, could form a layer on the ground, protecting the soil from the worst of the frost. Likewise, the first spring sun could penetrate down through the leafless canopy to warm the soils, stimulating the ground flora into rapid growth; only later, warmed through, did the deeper-rooting trees wake from their own sleep of hibernation, tapping deeper supplies of minerals from the soil. A new regime thus came into being, both within and above the soil.

The exact dates of these northward migrations are not of great importance, for they varied greatly along the north-south chequerboard of environments. An approximation of between 14,000 and 10,000 years before present saw the Taiga replaced by deciduous forest in the east and prairie grasslands in the west. The picture in the far west is complicated by the range of altitude offered by the Rocky Mountains and coastal ranges. There, the zonal shifts were more in respect of altitude than latitude, with the new vegetation belts climbing up ridges into hanging valleys on the flanks of the peaks. This new pattern of vegetation held sway until about 7,000 ago when the period of warming climate, the hypsithermal, reached its maximum.

A north-south series of nine peat-filled lakes in east Minnesota, covering a mere 200 miles, shows the pattern of this post-glacial vegetation to perfection. During the ice age, the whole area had been covered by the Rainy and Superior lobes of the continental ice sheet up until about 16,000 years ago. At the height of the hypsithermal the situation was as follows. Starting in the north, Weber Lake records a definite boundary between forests of Jack or Red Pine and those dominated by White Pine. The next three sites, Glatch Lake at plus 30, Spider Creek at plus 30, and Kotirania Lake plus 30 kilometres, each further south, record White Pine Forest. A mere 16 kilometres further south, Blackhoof records the border between White Pine and deciduous forest; while 8 kilometres further on, Jacobson and 16 kilometres beyond that, Rossburg Bog records deciduous forest. The next site, some 80 kilometres south, records Oak savannah, while Kirchner, 80 kilometres south again, was surrounded by true Prairie. This is how sensitive the pollen records can be in reflecting a climatic gradient over a small area. It is exciting to think that, as more and more of the peat, lake and alluvial sites are studied in sufficient detail, our knowledge of these dynamic changes will become more complete. It is also of great interest to speculate how far north and east the Prairie would have reached if the warm dry spell had continued unabated.

We know that it did not, for around that time the climate took a decisive turn for the worse and cooler, wetter conditions spread across the continent. However, by then man was well and truly present, adding his own influence, and that of fire, to the still changing scene. The great herds of megafauna had come and gone, browsing in the forests and grazing on the bounty of the prairies, perhaps helped into extinction by man the hunter. Could it be that their selective grazing had held back the spread of the broadleaf trees or aided the spread of grassland? So many questions will remain unanswered until the peatlands have been studied and understood in the detail they deserve.

The proof of climatic deterioration is, however, there written clear across the landscape, for about this time massive developments of new wetlands were initiated in areas as far removed as New England, Georgia, Florida and Minnesota, the most extensive being in the latter, where the flat bed of old Lake Agassiz, leached out by at least 3,000 years of rain, offered ideal conditions for the growth of Sphagnum, the best of the peat forming plants.

So, wherever one cares to look in America the vast majority of the soils and the vegetation are, geologically speaking, in their infancy, each having undergone a series of important and massive changes within the last 10,000 years. What is more, the peat-filled lakes have recorded the changes in great detail and, wherever they have been left intact, continue so to do.

Since the expansion of the wetlands which began about 4,500 years ago, the major event recorded in the peat is the coming of extensive agriculture as the white settlers pushed west, and from that point on there is no need to employ costly dating techniques, for their advance is part of written record. Starting about 1850, the peat records the gradual dilution of the pollen grains of deciduous trees and true prairie grasses by those of cereals and agricultural weeds, including Ambrosia, Chenopods, Iva and Xanthium. The fact that at this time the pollen of Corn makes its first real appearance on the prairie forest border supports the contention that the farm plots of the Indian, at least in these areas, were very small compared to those of the new agriculturalists.

Another marked feature of this time of change was the rapid infill of certain of the lakes with clay, presumably eroded from the surface layers of both the forest and prairie soils, which were then being re-opened after many thousands of years to the full erosive power of wind and rain. In Kentucky, limestone depressions have been found with as much as four metres of downwashed clay, all of which was rich in the pollen of Ambrosia and of newly imported species of weed and crop plants including wheat; a truly massive loss of soil. Shades of things to come! Perhaps the strangest of the imports brought from Europe and one that played havoc, especially with the wetter prairie soils, was our common garden

Earthworm. In comparison to the local members of the Annelida (the earthworm clan), *Lumbricus terrestris* was a voracious feeder and, once let loose, virtually destroyed the organic matter present in the deep soft prairie soils, undermining their evolved structure and speeding erosion.

The forest/prairie boundary had always been a good locale for human habitation, providing a variety of habitats with the promise of gatherable foods, a diversity of game both large and small and a handy supply of wood, for fuel and for construction. At first the white settlers shunned the prairies, for they thought no trees meant poor soil. However, once this misconception had been overcome, they too found that the boundary offered special benefits: ideal open space in which wheat and other crops could be planted without the need for felling trees and removing all the stumps. The first really large fields, unbordered by forest, thus came into existence and, with them, a whole new crop of problems appeared on the ever-widening agrarian horizons.

The settlers moved west for a variety of reasons, prime amongst which was the need to acquire new crop-land. Their own populations were on the increase as third, fourth and fifth generations went their own reproductive way. New immigrants came from all over the world, spurred on by wars, revolutions, pestilence and famine. Likewise, similar but more local factors pushed the farming families west to pastures new.

In 1778 the American Philosophical Society is said to have debated a number of problems which were troubling the farmers of New Jersey and New York. Amongst these were the ravages caused by the Angoumis Grain Moth and the Hessian Fly; the former reached the new, expanding, wheat belt from central America, the latter imported from Europe. The farming community responded in various ways. Some moved west; some gave up the unequal struggle and turned from arable to dairy farming; while others, like Jeremiah Wadsworth of Connecticut, stood their ground, took the matter seriously, and fought it. He asked his cousin to investigate a method of immunising wheat by steeping the seed in a solution of Elder. The findings of the study, which were positive, were reported to the *Connecticut Courant* shortly before the *American Mercury* announced that farmers had found a Bearded Wheat which resisted attack by the fly. In 1795 the same paper reported the results of a two-year trial of a new variety of Forward Wheat which had originated in Virginia. The fact that it matured some twenty days earlier than the other varieties meant it avoided the worst fly damage. So promising were these results that Wadsworth went into partnership with two local businessmen to import 2,500 bushels of this new wonder wheat for further development and dissemination. Their vision also stimulated the formation of a Society for the Promotion of Agriculture in the county of Hartford in 1797. Thus it was that necessity once more became the mother of invention, and the whole of society derived benefit from the new developments.

The depredations caused by the insect pests, whether imported or home grown, were, however, nothing when compared to the havoc which was being reaped by a much less motile organism, a parasitic fungus.

The blasting or rusting of wheat was first recorded in the New World in both Connecticut and Massachusetts in 1664 and, if you have ever seen a field of wheat suffering from this particular disease, you will understand how it got both its names. Black Stem Rust is caused by *Puccinia graminis*, a parasitic fungus which feeds upon the living tissues of the cereal, blackening and distorting its stem and greatly reducing its productivity. The parasite lives out only part of its life cycle on the crop; the rest and most important part, for it includes sexual reproduction, is carried out within the tissues of another plant, the Common Barberry. The sad thing is that before the white settlers came this plant wasn't common in America; in fact it wasn't there at all. If you look back at the list of plants grown by the Pilgrim Fathers you will see its name. They brought it to the New World for use in hedgerows, for its vicious three-pronged spines help provide a very effective barrier. What is more to the point, its straight light wood is ideal for making the handles of hand tools, its fruits are highly prized for use in sauces, wines, jellies and preserves; a good yellow dye can be obtained from its stems and various medicaments concocted from its parts. Altogether a very useful plant; unfortunately, its fungal parasite thought so too. In 1680, and again in 1685, days of New World prayer were ordained against the blast but all to no avail.

In Rouen in France a law had been passed as early as 1660 outlawing the cultivation of Barberry because the local farmers had noted that wheat was always blasted or rusted badly whenever it was grown in the vicinity of Barberry, the leaves of which were pock-marked with yellow scabs. Similar observations which related rust in wheat with the pollen of Barberry led to similar laws being passed in Connecticut in 1726, Massachusetts in 1754 and Rhode Island in 1766. It was not until 1865 that a German mycologist by the name of Du Barry proved beyond doubt that the spring rust of Barberry was caused by the same fungus which later in the year blasts the stem of Wheat, Oats, Barley, Rye and less useful grasses.

It has been estimated that an average-sized Barberry bush growing in Minnesota can bear about 35,000 leaves, 28,000 of which may be infected with the fungus. Inside each leaf grows the tiny threadlike body of the fungus, which is called a mycelium, sapping the energy of the living cells, until in early spring it breaks out onto the lower surface of the leaf. There is forms the yellow scablike structures which the farmers of Rouen rightly identified to be the trouble spots. If they could have seen them through a microscope, they would have seen that each was filled with aeciospores, tiny thick-walled structures which, once released, are carried away on the wind. An infected leaf can produce between 2.3 and 8 million spores, giving a total of some 64 billion per bush. Each spore is capable of infecting a cereal plant. Germination of the spores takes place on the surface of the new host plant, especially when there is a heavy morning dew. A new infective mycelium is thus formed and grows down through a stomatal pore into the living tissues beneath. After seven to ten days part of the mycelium grows back to the surface where it produces another type of fruiting body, a pustule containing as many as 200,000 uredospores. They are red in colour and each one can infect another cereal plant, which can produce another crop of uredospores within ten days. So

LIFE CYCLE OF RUST DISEASE

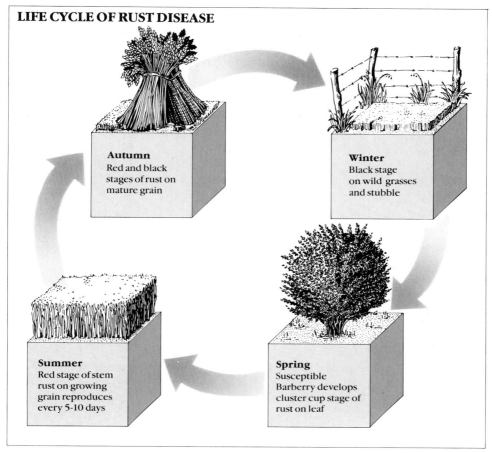

Autumn
Red and black
stages of rust on
mature grain

Winter
Black stage
on wild grasses
and stubble

Summer
Red stage of stem
rust on growing
grain reproduces
every 5-10 days

Spring
Susceptible
Barberry develops
cluster cup stage of
rust on leaf

it goes on, the disease running like wildfire through the crop, and the bigger and more continuous the fields, the faster will it spread. By harvest time, so many spores can be present in the crop that great rusty-red clouds rise up before the scythe, as the reapers move through. By then the damage has already been done, and unfortunately this is not the end of the story. Before the crop matures, the fungus switches over to the production of a third type of spore; they are black and are called teleospores. These spores have very thick walls which protect their living contents so effectively that they can overwinter, germinating next spring, especially during wet weather. Each teleospore consists of two living cells which produce short jointed mycelia on which are borne four tiny basidiospores. These are shot off again to be carried away by the wind, and they can infect a Barberry, thus starting the whole cycle once more.

The fascinating thing is that of all the types of spores produced, only the black teliospores can overwinter in the cooler areas north of Texas, and what is more, they cannot re-infect the cereals without first completing their life cycle inside the Barberry. Re-infection must therefore seem a pretty long shot, but all it needs is one infected Barberry bush to put the fungus back in business. The most infamous example was when after the bumper wheat crop of 1915, there followed a season which started off both cool and moist. The farmers of the Great Plains and Prairie Provinces didn't worry too much, because eventually their crop got off to a late but lush start. The local Barberry bushes were already heavily infested. Hot humid weather from May through June favoured both spore production and germination, and by July stem rust was widespread throughout the Spring Wheat area. August was hot and dry and premature ripening brought disaster.

In 1916, 14 million acres of wheat were planted in Minnesota and North and South Dakota; yield was down to 5.5 to 7.6 bushels per acre, only a third of that of the previous year. An estimated 200 million bushels of wheat were lost in America and a further 100 million in Canada: a loss of 480 million dollars to the farmers and a blow to the economy of the whole world, which was then in the grip of World War I. Flour prices climbed from $6.10 to $8.60 per barrel, bread queues in London grew longer and longer and people went hungry, for there was no official rationing in force. Something had to be done to ensure adequate food for the Allies. After many meetings and conferences a national programme to control Stem Rust was announced, part of which centred on the eradication of Barberry from the wheat belt.

A food queue in London during World War I.

That is why, in 1918, our farmer friend in Crystal Bay uprooted his hedge and raised his first successful crop of Oats for ten years. Total eradication in one area, even if it were possible, is not, however, the complete answer, for red uredospores arriving from further south, where they can overwinter, cause an estimated one infection per 14 metres of planted cereal, that is 1,500 infections per acre. This could in turn produce 300 million spores; a prodigious number, but only 0.5% of those which could originate from one infected Barberry bush, and a medium-sized one at that.

However, much more insidious is the fact that it is only in the Barberry stage that the fungus can complete the most important phase of its life cycle. Before the tiny aeciospores can be produced, sexual reproduction must take place, and as in all advanced organisms, this brings about a re-shuffling of the genetic material by recombination of that derived from the two parental stocks. Thus new genetic variety comes into being.

Wheat, Oats and Rye are each attacked by different varieties of Stem Rust; Wheat and Barley are attacked by the same one. Each rust variety is made up of many different races, their variety maintained by sexual recombination and mutation.

To date, more than 200 different races of Stem Rust have been identified on Wheat alone. The races differ, both in their virulence and their ability to infect different varieties of grain. A particular variety of cereal may appear to be immune for many years, until a new strain of fungus blows in from a nearby Barberry.

The removal of the Barberry host thus cuts down the possibility of variation in the fungus, ensuring that fewer varieties will be present in an area in subsequent crops and (perhaps most important) increasing the likelihood that the same variety of fungus will be common from year to year. The plant breeder and the farmer are thus given a chance, the former to develop more resistant varieties of cereal, the latter to plant and propagate the same immune crop from year to year.

Imagine the problems of a plant breeder who must evaluate his or her new cultivars against 28 different races of fungus each year, in the expectation that six of them will never appear again and that at least three new ones will come up the following year. This was indeed the situation in 1918/1919, which saw the end of one sort of war and the start of another.

Since that time, 500 million Barberry bushes have been destroyed in the USA alone, with an estimated saving to the farmers of $300 million per annum. What is more, there is now sound evidence that across the whole of the northern Wheat Belt the fungus, now estranged from its alternative host, is so weakened that it is less able to produce teleospores, and those that are produced are losing their ability to germinate and re-infect the Barberry.

Now that we know, it is so easy to understand the importance of the laws, but the original observations by the farmers and their insistence in the face of disbelief and even derision must be given all the credit it deserves—a real stroke of home-spun genius. The fight, however, still goes on and vigil must constantly be kept in order that the plant breeders remain one jump ahead of the fungus.

It was the early success of the eradication programme; the diamond-tipped drills cutting their way through to what for a long time appeared to be a limitless supply of fossil fuel; the success of the production lines of Henry T. Ford; and the cessation of hostilities in Europe, that led to the real opening up of the Wheat Belt as the world has come to know and depend on it.

The war, though totally catastrophic in humanitarian terms, hardly made a dent in the human population curve. Peace, and the post-war baby boom, coupled with emigration, meant more mouths to feed in the New World and an expanding market to supply with flour power. There was land, lots of land, and sunny skies above, and with much of the indigenous wild life now gone, there was not so much need to fence the farmland in, so agricultural production bloomed on all fronts to meet the new demands. Tractors, driven by horsepower rather than towed by horses, made hay redundant as they began to rip into virgin soils and dig deeper into those already opened up. Most favoured were the soft Mollisols which had developed along with the Prairie Grasslands and Moss-rich Forests over many thousands of changing years, each to its own productive maturity. The gentle processes of chemical and biological weathering had been working away at the parent material throughout that time, keeping up the supply of nutrients necessary to top up the natural process of recycling which maintained annual growth.

Motorised farm machinery brought a revolution in intensive agriculture.

All this came to an abrupt end as the soils were opened up to the brashness of mechanical weathering (no, erosion) by ploughshares, rain and wind, the severity of the latter two depending on the vagaries of local climate.

Thus it was that a new menace began to take grip, at first unseen, on the farmscapes, eating their goodness away. All the time the new fields were surrounded by forest or natural·grassland, such effects were localised by their binding, stabilising presence, but as more of each landscape was put to the plough, the problem spread, erosion by both water and wind becoming a real menace.

Suddenly, in the 1930s, the whole world knew that something had gone wrong. Drought across Texas and Oklahoma lifted the upper layers of the now dry soils into huge rolling clouds of dust, which obliterated the sun as they passed on their way. The dust contained much of their clay and organic matter, which had held the nutrients in available form and supported, at the most, fifty harvests. In just

Bitter harvest of intensive agriculture: the 'dust bowl' and bankrupt farmers.

a few years almost 300 million acres of farmland were severely damaged, and bankrupt farmers Model T'd it to the west. The good earth had become no more than dirt; the stuff which had supported their dreams just upped and flew away. The fruits of their labours were gone, yet the grapes of their wrath came to fruition when President Franklin Delano Roosevelt enacted new laws, and created the United States Soil Conservation Service.

Its aim was to put heart back into the soils, and it worked. With government agents to preach and teach the know-how of 'hold on to our soil, it's the most precious thing we've got', and money to help put the advice into operation, terraces were carved out, windbreaks planted, as contour ploughing, catchment management, strip cropping and crop rotation became more and more the order of the farming way. Soil conservation became the 'in thing' of after-dinner discussions, school camp curricula and cub-scout badges as the nation took a firmer grip on

An answer to soil erosion: strip cropping and contour ploughing.

its soils, a rich resource, which had been handed to them on a diamond-edged glacial plate. Once more the all-American soils began to bear golden fruit from coast to coast, and the seas became shining again, untainted by the telltale slicks of mineral-rich clay which had, for years, flocculated with each outgoing tide at the mouths of many of America's rivers.

Over the next forty years both the soils and the farmers gained new heart, their pacemakers being the development of new farm implements, crop varieties, chemical fertilisers, pesticides and herbicides. So effective was it all that every year the hungry world came more and more to depend on the great American Agricultural Revolution.

Worldwide medicare, backed by new drugs, mass immunisation and antibiotics, many developed under the impetus of war, cut into human mortality and primed the population bomb. The concept of a Third World hardened into the reality of hungry eyes, which looked towards the World's Best Friend and pleaded for a slice of the American loaf.

The agricultural real estate had, by this time, become so productive that legislation had to be enacted and farmers compensated to keep land out of production, thereby avoiding gluts of produce and the irony of non-economic farming. At the same time charity, in the guise of keeping up with opposing ideologies, emptied some of the great storage silos and filled hungry bellies with hand-out hope.

About this time, three events of great importance rocked the world in rapid succession. The Green Revolution, spearheaded by plant breeders and agronomists, produced the Super Cereals and the hope for a well-fed world. Agricultural aspirations of two countries which, though less than half a world away, were ideologically light years apart, crashed in the face of drought and collective malpractice. This opened up new links, both with eastern Asia and across the Bering Straits. Catastrophic grain harvests forced both Russia and China to buy wheat from the New World. The farmers, in both Canada and America, responded to this new lucrative demand and it is not an unfair thing to say that they laid aside many of the soil conservation practices that had been so hard won and so successful since the 1930s.

It would be both trite and wrong to blame the farmers and say that it was short-term gain which alone motivated their new endeavours, for at about this time the third of the great world events took place.

Suddenly, the main producers of the fossil fuels came to realise the full value and power of their non-renewable resource. Nature had taken more than 200 million years to lay down the world stocks of oil; man had used up over half of this rich store in a little over 100 years of living, riotous at least for some. All the time the world's economy was in buoyant mood, money made from oil could be invested in other ways which guaranteed good return; you could, in fact, 'Sheik' your money and take your chance on the stock markets. However, once recession had set in, it was deemed by some to be more prudent to raise the price of oil and leave it in the ground until things looked up! That one exacerbated the other is of little short-term account, at least for those countries whose riches are founded only on oil. The problem was, and still is, that in terms of useful energy obtained, modern agricultural practices are uneconomic: they have themselves become hungry for energy. Horses eat hay, a renewable resource; the more horses you use, the more hay must be produced and the more organic manure is released for use. 'Horsepower' needs gas, and the more work your motor does the more gas it guzzles—and horsepower exhaust is nowhere as useful as the good old variety. Worst of all in this 'you can't get something for nothing' struggle for power, all the new farm chemicals on which agriculture now depends, and without which its productivity would slump, require energy for their manufacture, transport and application. What is more, the raw material from which many of them are made is crude oil.

Balancing the energy equation becomes even more critical as nutrients lost by erosion must be replaced by the all-purpose fertiliser: potassium, nitrogen and phosphorus, KNP for short. The 'N' in this world club sandwich requires enormous amounts of energy to get it into active form. Nitrogen is the most abundant constituent of the earth's atmosphere, making up 78% of its volume, but in that form it is inert. A miracle of early evolution was simple plants, bacteria and their kin, which could fix this elemental nitrogen into nitrate, in which form it can be used by other plants. A miracle of chemical engineering has allowed the chemical giants to do the same thing, at a price, and that goes up with the price of energy.

That is why the Green Revolution failed to answer the problem of the Third World. Super Cereals are indeed super-productive, but super productivity requires super amounts of super-fertiliser, including nitrogen.

The world thus had gas at a price, and certain sections of it had either the credit or the creditability to pay for it. So what can now only be called the American Agricultural Machine rolled into action, and to keep the energy equation as nearly balanced as possible (for, remember, energy can neither be created nor destroyed) production costs were cut to the bone, and the soils, once again, began to bear the brunt of short-term aims. Who can blame them, for the stark statistics state that in 1910 each joule of energy produced by the American Agricultural Machine had required one joule for its production; by the 1970s this figure had risen to an uneconomic ten joules.

As a convenient rule of green fingers, the Soil Conservation Service has calculated that an average soil is formed at the rate of five tonnes per acre per year. This may seem a lot, but spread out over that area it amounts to, at the most, a few precious millimetres of parent material transformed into living soil. The argument, there-fore, goes something like this: if 5 tonnes are formed per acre per year, then 5 tonnes can be lost without causing too much concern.

In 1977, according to the Soil Conservation Service report, 2,000,000,000 (yes, two billion) tonnes of soil were lost through sheet and rill erosion. Critical losses were recorded in the Corn Belt, Delta States and west Tennessee, while gully ero-sion stole another 450 million tonnes and streambank and roadside erosion even more. With record losses measured at between 50 and 100 tonnes per acre in the Palouse area of Washington, Oregon and Idaho, and most extensive losses in the critically important Corn Belt, with average losses in Iowa of 9.9, Illinois of 6.7 and Missouri of 10.7 tonnes per acre per annum, the future looks black for both the farmers and the hungry world.

Add to this tale of woe the hard fact that it is not just the losses of topsoil that are causing concern, but that wherever it goes it causes pollution. In situ, it held the mineral nutrients in cycle and maintained the hanging reservoir of water, which buffered both against drought, flash floods and the rapid enrichment of ground water with both nutrients and pollutants. Once the structured soil is on the move the incidence of flooding increases; ditches, which require energy to dig and keep clear, become choked with silt; as do streams and rivers. Fisheries and amenities are lost as chemical pollution of many types spreads with the clinging tide of clay-rich mud. Agriculture has thus become the single biggest polluter of two-thirds of the river basins in the USA.

At some time all this will have to be stopped, and this will cost energy and big, big money—and all the time, the world's stock of fossil fuel is running down and the value of what remains is going up.

To keep in business many farmers have already cut back on the production of primary food crops, and turned to more lucrative cash crops. They grow Rape and Sunflowers for their seeds and the oil that can be extracted from them, Soya Bean for animal food and textured plant protein, and more and more corn

THE CHANGING FACE OF THE USA

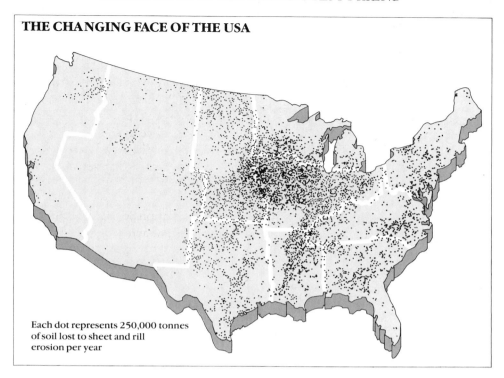

Each dot represents 250,000 tonnes of soil lost to sheet and rill erosion per year

for purposes other than human food. Some have given up the uneconomic struggle entirely and gone into the investment business, for each year a staggering three million acres of rural land are lost to non-agricultural purposes. Much of that figure represents good cropland (the figures that have illuminated the location maps at the head of each chapter). Good cropland is level, well-drained land which can be cropped year after year with the least damage to the soil. These same attributes, however, make it ideal for many other uses: housing, highways, airports, parking-lots and even for the construction of amenity lakes and reservoirs to bolster the local water supply and help keep both urban and rural taps flowing.

The message is written clear across the face of America: the once self-regenerating soils of these broad acres, which were handed to man on the glacial plate, cannot be expected forever to feed an ever-increasing proportion of the human population: a bitter message for a world in which twenty-eight children under the age of five die of conditions related to malnutrition every minute of every day—a message that has been spelt out again and again in this account of America's natural heritage.

CHAPTER TWELVE

The Declaration of Dependence

I N 1776 Thomas Jefferson, a farmer by upbringing and a statesman by nature, drafted The American Declaration of Independence. There can be no better amalgam of attributes than these, for a good farmer keeps his lands in good heart, and a good statesman gives his whole heart to his lands, and there is, to date, no better statement of the aspirations of humanity.

The declaration begins like this:

When in the course of human events it becomes necessary for one people to dissolve their political bonds with another and to assume among the powers of earth, the separate and equal station to which the laws of nature and of nature's god entitle them, a decent respect to the opinions of mankind requires that they should declare the causes which impel them to the separation.

We hold these truths to be self evident, that all men are created equal, that they are endowed by their Creator with certain inalienable rights, that among these are life, liberty and the pursuit of happiness.

'The laws of nature and of nature's god' have throughout this natural history spelt out the stark message of limitation.

The message has been the same, from the stepping-stone world of the Aleutians, the green islands whose biological diversity is strictly limited by area, down to the awesome reality of the Great American Interchange, when the flora and fauna of two long-separated continents came into conflict for limited resources, and only the fittest survived. The potential of any land mass to support life is limited, at least in the long term, and the development and maintenance of that potential depends on a diversity of life.

North America's rich pre-history has expanded on this story of limitation, showing that man is part of the continuity of living creatures, and hence is subject to the same rules of limitation.

The first Americans followed the animals of the chase into the New World with no knowledge of the richness of opportunity that lay ahead. They were the true pioneers, with no maps to guide them on their way, no concept of a planet, flat or round, to limit their endeavours. The next horizon was the limit to their nomadic vision, and the bounty of the land between determined the reality of success or

The statue of Thomas Jefferson at Washington, D.C.

failure. What was it, then, that spurred them on to colonise the diversity of two whole continents in a mere few thousand years?

One can only suggest that competition between the needs and greeds of their own burgeoning populations pushed certain factions among them ever onwards.

The struggle for existence always brings out the three traits inherent in any population. First there are those individuals best called the Hawks, who will always stand up and fight, whatever the opposition, and who will according to the laws of chance have no more than a fifty-fifty chance of winning the battle for livelihood. Then there are those best looked upon as Doves who will acquiesce in every situation and if given the opportunity will turn and run away and hence may live to turn another day. These too will enjoy only an even chance of survival. Last but by no means least there are the in-betweeners, the floating voters who size up each and every situation and only then decide to stay and fight or turn and run. Only these stand a better than even chance of winning the battle for life; they must therefore be regarded as the fittest, the long-term winners in the struggle for existence.

With all the diverse richness of two continents at their disposal all three traits within the human make-up could be accommodated and thus encouraged; all territories were soon annexed as man exploded across the New World. Wherever there was potential, from Ellesmere Island in the far north to Tierra del Fuego in the south, it was soon put to human use, the special attributes of each area bringing out the special skills of *Homo sapiens*. This was not the blind chance and necessity which had marked the slow progress of earlier stages of evolution but the positive seeing and seizing of each and every opportunity; the realisation of what was necessary, and the doing of it. Special skills of hunting, fishing and the gathering and preparation of roots, shoots, flowers and fruits, each in their own season, all fostered, and were fostered by, dependence on more local knowledge. Settlement rather than nomadism, a scaling down of horizons and the need for defence against hawkish insurgence, structured new ways of life.

Only time and further excavation will tell the full story in the detail it deserves and raise sites like Cape Kreusenstern, Onion Portage, Serpent Mound, Newark Earthwork, Cahokia, Keet Seel, Canyon de Chelly and many many more into the world list of heritage sites which are a must of every grand tour the aim of which is to gain a fuller understanding of this thing called human kind.

Already the evidence states beyond all doubt that once the Bering Straits had closed behind the first Americans there was no need for any contact with any other world, old or extra-terrestrial. Many of the main breakthroughs of the technology of human culture happened here. Likewise, despite any barriers of regional distinction which came into being, each breakthrough in New World culture spread with all speed through the twinned continents.

The craft of knapping flints; of fashioning and attaching projectile points; the strategies of hunting game, both big and small; the necessity and use of the atlatl as the game become smaller; the use of baskets to aid collection, and pottery to improve the techniques of storage and cooking and hence the digestibility and energy content of that which had been collected: each step along the line simplified the problems of existence, putting a new resource, time, at the disposal of dexterous hands and thinking minds.

That the forests of the Eastern Connection, themselves mere remnants of a once world-wide sylvan culture, were rich enough to foster the development of a human society, itself enriched with civil arts, crafts, religion, engineering and science, all based on a hunting and gathering economy, is of great interest and importance. It seems reasonable to argue that the Adena and Hopwellian cultures represent the zenith of man's development as part of a natural system and that somewhere amongst those now scrub-covered mounds evidence will be found relating to their increasing dependence on crops and to the reason for their demise.

The domestication of corn and its development as a crop which could be grown from the Gulf of St. Lawrence in the north to central Chile in the south stands as the greatest achievement of the New World, for it made possible great cultural developments. The Hohokam, Mogollon, Patayan, Anasazi and Mississippian cultures stand alongside those much better known from mid- and meso-America and of the Old World. Corn formed the keystone crop of a simple polyculture which provided men with a balanced diet, the energy and the body-building power to lift their populations and their achievements well above the natural carrying capacity of their landscapes. They came, held their glorious sways and passed into pre-history. Their lands, reverting to nature, provided a diversity of products to be hunted and gathered by less demanding peoples with an appropriately less demanding way of life.

All this and much more took place within the boundaries of North America, that section of the super-continent Pangaea which was above all others blessed with a size, shape, topography and line of drift which placed the world's richest soils in a temperate well-watered climate, into the path of an ice age, and soon after gave them into the hands of man.

So it was when history was thrust upon the New World with the coming of European influence.

Pre-history by its nature cannot tell lies, though its artefacts may lead the unwary to ill-founded conclusions. History, on the other hand, is recorded. The record is influenced by all the inconsistencies of human nature, and so can pull the white wool of untruth over the eyes of even the most discerning thinker. The usurpation of one people's lands by another, the creation of massive new opportunity, the founding of the most affluent nation on earth: where were and are the rights and wrongs? Who can judge? All one can do is take council of the richness of that history, look with compassion upon all colours of its arguments and put that compassion to good purpose.

'All men are created equal—they are endowed with certain inalienable rights—life, liberty and the pursuit of happiness.' A highlight of any grand tour of America must be to stand beside Jefferson's statue on Capitol Hill, Washington, there to think about those words. I defy any human being not to be moved by their meaning and by the powerful presence of the man Thomas Jefferson. His image stands on a plinth decorated with *Zea mays*, the plant which above all others changed and is still changing the face of America.

In 1776 the vast majority of the soils of the great continent of North America were still in their natural state, their living systems intact. Two hundred years later the American Dream had become the reality of the most affluent nation on earth, an affluence built on 170 million hectares of soil under intense cultivation, 34 million producing corn, 30 million wheat, 25 million hay, 24 million soya bean. All except hay are major exports, key to both the American economy and to a hungry world, and all are increasingly dependent on the massive use of fossil fuels, a situation which cannot continue for much longer.

Corn, the amazing miracle of America's pre-history, is still the number one crop, but now produces much more than mere food for the multinations. Its many products are vital resources for much of the industry which supports this rich country, its aspirations and inspirations, both at home and abroad. To indicate but a few: corn products are used in the manufacture of paper, paints, plastics, textiles, sweets, cosmetics, car tyres, gum, glue, soap, dynamite, fluxes used in the manufacture of metals and lubricants used in the drilling of oil wells. It is true to say that if the corn crop failed, America and its economy could begin to fall to pieces. What is more, science and commerce are working together all the time to squeeze more and more out of this versatile crop. It is not simply a dream of the Corn Barons that much of the automotive power which moves America in air-conditioned comfort will soon be supplied by Gasohol made from corn. But at what cost to a starving world and to those now hungry soils?

To meet these growing needs the plant breeders, agronomists, agricultural engineers and 'farmaceutical' manufacturers have concentrated much of their attention on *Zea mays*, making it one of the most pampered and uniform mega-crops on earth. Where in 1776 local Indians and colonists planted many different varieties of corn, each with its own characteristics of growth—tolerance to flood, wind, drought, disease and all combinations of local conditions—their modern counterparts, a few farming families, cultivate multi-millions of absolutely uniform plants. Each plant is as alike as an identical twin, each bred for performance and productivity under the norm conditions which statistics prove should prevail in its area of cultivation.

One recent breakthrough in this continual process of upbreeding was the discovery of what has come to be known in the hybrid corn trade as Texas Cytoplasm. This has the unique property of conferring impotency in the guise of male sterility upon its progeny: in nature, a genetic formula for rapid extinction, but in the hands of the plant breeders a unique tool, for it does away with the time-consuming, labour-intensive and hence costly need to detassel the plants to avoid the crossing of wrong lines through self-fertilisation. So successful was this new innovation that by 1969 more than 15% of the total American crop boasted cytoplasm from the Lone Star State.

That same year a virulent mutant fungus, which causes southern corn blight, cropped up in Illinois. Later in 1969 it was causing havoc among the winter crops of Florida, and the damp season (well outside the statistical norm) of 1970 brought instant disaster: 15% of the nation's corn crop succumbed to the disease—the 15% that contained Texas Cytoplasm in its make-up.

Thanks to the immense fund of knowledge concerning the genetics of corn, and the heroic efforts of the breeders, the situation was saved and by 1973 there was sufficient new hybrid corn unconnected with the Texan line to keep the farmers growing. This was opportune, for 1972 had seen a decline in the global production of food for the first time since World War II, with disastrous harvests forcing both China and Russia to buy thirty million tonnes of cereal from North America. Through good fortune, then, the corn crisis passed with nothing more disturbing than an increase in the price of certain foods. The shock to the agricultural system, however, went much deeper. It brought about the realisation that America's agricultural economy was becoming more and more critically balanced, not only on the knife-edge of costs against returns (energy in, versus energy out) but on an ever-narrowing pinpoint of genetic uniformity.

To ram the point home, 1976, the Bicentenary of the Declaration of Independence, saw the start of a chain of events which led to a discovery of great importance.

Hugh Iltis, champion of the Teosinte-primitive corn hybrid theory of the origin of *Zea mays*, sent a New Year's card to all his colleagues. On it was depicted *Zea perennis*. One of these cards was received by Maria Luz Puga, Professor of Botany at the University of Guadalajara in Mexico. He challenged his undergraduates to find the mysterious Perennial Teosinte which since its discovery in Mexico in 1910 had only been seen occasionally and was now thought to be extinct in the wild. One student, Raphael Guzman, accepted the challenge and after an arduous journey backpacking through the Sierra de Mantalan found what turned out to be much more than he was looking for: a plant whose above-ground parts looked not unlike Annual Teosinte, but which below ground possessed a rhizome proving beyond doubt that it was a perennial.

Think of the benefits of a perennial corn crop. If only this unique property could be bred into corn itself. Unfortunately it was known that this was impossible, for *Zea perennis* has forty chromosomes in each cell, twice the number of *Zea mays*, a difference which makes the two incompatible, hence the two specific names.

Nevertheless, Guzman appeared to have discovered the elusive *Z. perennis* in the wild and soon seeds sent to Hugh Iltis were growing in the Botanic Garden at Madison. Imagine his excitement when he took his first look at the plants. Imagine his surprise when he realised that they were not *Zea perennis*, nothing resembling the dried specimens he had seen in the Smithsonian collections. A chromosome count confirmed his suspicions. The plants were indeed different; they had only 20 chromosomes. A completely new plant which deserved a separate status and a new name: *Zea diploperennis*. It was then that the real significance of the find dawned on Iltis. Here was what must be the most primitive ancestor of corn discovered to date, a direct ancestor in the line of ascent, and if it could be crossed with *Zea mays* it could be the key to a whole new era of corn affluence. No need to deal with the underground remains of last year's crop; no need to re-plough, opening up the soil to the problems of erosion; no need to re-seed. All that time, effort and, above all, energy, could be saved. What is more, such a primitive ancestor could well contain genetic information relating to a broad spectrum of disease-resistance and habitat-tolerance which had been lost from the more advanced inbred lines. The fact that the new plant was found growing on the edge of a stream on a slope facing north-east, 2,700 metres up some very misty mountains, pointed to all sorts of new cropabilities.

However, the most significant point to my story is that the plant itself was restricted to a mere two hectares of land where it grew as a weed between the farmer's conventional fields of corn. He tolerated its presence because, as he said through an interpreter, the rock-hard Teosinte could be used as a food in times of real famine. In times of real famine! It is almost unthinkable: one agricultural development grant in the wrong place, a new super-productive hybrid corn to plant, a back hoe for a day to level and improve the drainage of the fields, and *Zea diploperennis*, a potential multi-million dollar resource, could have gone forever—and that farmer and his family would have had nothing to fall back on when the next inevitable famine came.

This chain of events has brought about a radical change in attitude. Discussions have begun concerning the importance of the conservation of genetic resources, not among the long-haired idiot-fringe of environmentalists and ecologists, but among the hardest of hard-headed businessmen. They are asking, where is the genetic diversity which has served their industry (and the New World) so well, for so long, and which is the key to the well-being of its future? Where is the real heritage of the American way of life?

The answer is there, staring them in the face. Much has already disappeared; a miserable pittance is lodged in the semi-safety of refrigerated concrete bunkers, gene banks at a few plant-breeding institutes. The vast majority of what little remains is in the outback fields of Indian reservations and the less developed parts of Meso and South America. Only there do the people farm in the time-honoured way, sowing a little of each harvest to form the basis of the next crop. Only they maintain the diverse purity of each line, conserving the many unique genetic messages of survival. If action isn't taken now, it will be lost forever.

The same is true, not only for all the other major crops which now support man, but for the many new crops which must be developed in the not too far distant future: crops which will provide the raw materials for future industry.

When we have used up all the coal, oil and natural gas—what then? It must be remembered that they are not only sources of energy, but are the raw materials for much of the chemical and plastics industries. Even if science can find an alternative, safe, source of energy—and I believe it can—where are the raw materials for our future chemo-plastic way of life? The answer lies in the genetic diversity of the world, those four million other sorts of living organism with which we share this planet. There is a vast renewable resource there for the taking, but again time is desperately short. Clear across the first, second and third worlds, development, couched in terms of progress, is sweeping away the diverse inheritance of evolution, replacing it with an ever-increasing uniformity of crops, all of which require care, and massive use of energy, before, during and after germination. The world is losing 70 hectares of forest every minute. Of the 250,000 species of flower-bearing plants known to exist, 25,000 are on the brink of extinction, and more are following the dinosaurs, Dodos, Stellar's Sea Cows and Passenger Pigeons into limbo every day. The world could lose one million species of organised life by the end of this century. The stuff of evolution, genetic diversity, is being drastically reduced. The survival pack of this green earth, the age-old information which was held in store against fire, flood, drought, earthquake, hurricane, ice age and more subtle environmental change is being destroyed in the name of progress.

The subsistence fields, waste places, wildernesses, National Parks and Nature Reserves of the world thus take on a new and vital role. They can no longer be regarded as anachronisms in the 20th century to be swept away by short-term grant aid and fast megabuck development. Each one is part of a genetic storehouse, a unique investment of immense value to the future.

Homo sapiens, the pioneer days are over. The wilderness has been tamed, and the little that remains is no longer a threat to our existence, it is the cornerstone of any hope we have for the future.

This is the end of the American dream, and yet it could be the beginning of another of much greater and more lasting consequence, a new dream—no, a new world reality—still firmly rooted in Jefferson's words: 'the laws of nature and of nature's god'; 'all men are created equal'; and in the soils of this great country.

This whole natural history has been a declaration of dependence on the laws of nature; laws which lay down a statute of limitation, laws which demand equality for all living things, for within the diversity of life each organism has an important role in the economy of nature. The future of humankind lies safe only in so far as we take heed of these laws.

Can it be done? Can this new dream come to pass before the imperatives of struggle for diminishing, non-renewable resources condemn the world to holocaust?

The answer is yes—if we act now. All that is needed is a change in attitude. The hawks and the doves have had their chance. Now the time of the in-betweeners, the thinking, floating voters, is here, and they must stand up and fight for their inalienable rights.

And it is happening, even in the most affluent region of this the most affluent nation of earth. Remember Mono Lake.

Further Reading

The literature on American natural history, pre-history and history is immense, and to give an adequate introduction to this vast store of knowledge is almost impossible. Below is listed some of the literature which has been of greatest use to me in developing this book and the films which illustrate its narrative.

AGRICULTURE

JONES, E. L. *Creative disruptions in American agriculture, 1620–1820*. *Agricultural History* XLVII, 1974 pp 510–528.

LEACH, Gerald *Energy and food production* I.P.C. Science and Technology Press, ne, 1976

UNITED STATES Dept. of Agriculture, Soil Conservation Service *America's soil and water conditions and trends* Syosset, NY: Water Information Center, 1980.

FAUNA

KURTEN, Bjorn and ANDERSON, Elaine *Pleistocene mammals of North America* Columbia University Press, 1980.

NITECKI, Matthew H. *Biotic crises in ecological and evolutionary time* Academic Press, 1981.

FLORA

HOLT, Perry C. and others. *The distributional history of the biota of the Southern Appalachians* 3 vols. University Press of Virginia, 1969–77.

HULTEN, Eric *Flora of Alaska and neighboring territories: a manual of the vascular plants* Stanford University Press, 1968.

RAVEN, P. H. and AXELROD, D. I. *Biogeography and past continental movements* in *Annals of the Missouri Botanical Garden* vol. 61, no. 3, 1974.

RAVEN, P. H. and AXELROD, D. I. *Origin and relationships of the Californian flora* (Publications in Botany, 72) University of California Press, 1978.

GENERAL

Arizona Highways Magazine, Vols. 1–58. Phoenix, AZ. monthly.

WALTER, Heinrich *Vegetation of the earth* (Heidelberg Science Library, 15) New York: Springer-Verlag, ne, 1979.

WHITTAKER, Robert H. *Communities and ecosystems* New York: Macmillan Publishing Co., ne, 1975; Collier-Macmillan, paperback, 1975.

LOCATIONS

SUTTON, Ann and Myron *Wilderness areas of North America* New York: Funk and Wagnalls, 1974.

WILSON, Josleen *The passionate amateur's guide to archaeology in the United States* (Collier Books) New York: Macmillan, 1981.

PREHISTORY/HISTORY

BRENNAN, Louis A. *American dawn: a new model of American pre-history* London: Macmillan, 1970.

DRIVER, Harold E. *Indians of North America* University of Chicago Press, ne, 1969.

'SCIENTIFIC AMERICAN' *Avenues to antiquity: readings from 'Scientific American'* introduced by B. M. Fagan. San Francisco: W. H. Freeman, 1976.

STEWART, Hilary *Indian fishing: early methods of the Northwest Coast* University of Washington Press, 1977; J. J. Douglas, Canada, 1978.

TURNER, N. J. *Food plants of British Columbia Indians* 2 parts. (Publications 34 and 36) British Columbia Provincial Museum, 1975–77.

TURNER, N. J. *Plants in British Columbia Indian technology* (Publication 38) British Columbia Provincial Museum, 1979.

SOIL

JENNY, Hans *The soil resource: origin and behavior* (Ecological studies, 37) New York: Springer-Verlag, 1981.

Acknowledgements

My thanks are due to Dr Hans Jenny, University of California, Berkeley; Dr Peter Raven, Missouri Botanical Garden, St Louis; Dr Daniel Axelrod, University of California, Davis.

Ms Patricia Allen, University of California; Dr Cliff Amundsen, University of Tennessee, Knoxville; Dr Douglas Anderson, Brown University, Rhode Island; Dr James Anderson, Cahokia Mount National Monument; Ms Nancy Ash, Kampsville Archaeology Centre, Illinois; Bradley Baker, Ohio Historical Centre; Dr James Baker, Plymouth Plantation, Massachusetts; Dr Julio Betancourt, University of Arizona, Tucson; Jim Bilyeu, Soil Conservation Service, Nashville, Tennessee; Dr Robert Bird, Institute for Study of Plants, Food and Man, Missouri; Dr Vorsilla Bohrer, Eastern New Mexico University; Dr David Boufford, Grey Herbarium, Cambridge, Massachusetts; Mr Gregory Byrd, Racho la Brea; Ms Kathleen Byrd, University of Louisiana; Dr Jack Carter, North Dakota State University, Fargo; Dr and Mrs John Chandler, Los Angeles; Dr Edward Clebsh, University of Tennessee, Knoxville; Mr Larry Cole, Massachusetts; Mr Richard Countryman; Dr Edward Cushing, University of Minnesota, Minneapolis; Ms Leslie Dawson, California State Parks; Mrs Mary de Decker; Carl Deu, Savage Garden, Lake City, Tennessee; Dr Harvey Doner, University of California, Berkeley; Craig Engel, Calaveras Big Trees State Park; Lawrence Erickson, Ounalashka Corporation; Mrs Susan Faulkner, E.I.L.; Dr Paul Fish, University of Arizona, Tucson; Dr Richard Ford, University of Michigan, Ann Arbor; Steven Frowine, Missouri Botanical Garden; Dr David Gaines, Mono Lake Committee; Miss Duane Garrison, Tiffany's, New York; Dr and Mrs Nicholas Gessler, Queen Charlotte Island Museum, British Columbia; Ms Florence Givens, University of Louisiana; Ms Kathleen Groodie, Museum of New Mexico, Santa Fé; Dr Eric Grimm, University of Minnesota, Minneapolis; Haido Nation, Queen Charlotte Islands; Dr Heinrich Harries, Mount Allison University, Sackville, New Brunswick; Mr Michael Helbling, Moscow, Idaho; Professor Michael and Mrs Peggy Hoffman, University of Arkansas; Mrs Judith Ingrams, Plymouth Plantation, Massachusetts; Ironwood Country Club, Palm Desert, California; Mr Russell Irvine, British Columbia Heritage Trust; Mr Barry L. Jenkins, US Department of Agriculture, Washington, DC; Jack Jenkins, Moscow, Idaho; Mr Myles Johnson, Clay County Fair, Iowa; Professor E. L. Jones, University of Western Australia; Mr William Knight, US Department of the Interior, Bureau of Land Management, New Mexico; Mr John Larsen, Cooperative Extension, University of California; Mr John C. Larsen, US Forest Service, Calaveras; Mr Joe Larssen, US Soil Conservation Office, Washington; Dr Austin Long, University of Arizona, Tucson; David Long, University of Minnesota, St Paul; Mr William McClung, Lake Minerals Corporation; Dr Raymond McCord, University of Tennessee, Knoxville; Mrs Lillie McGarvey, Unalaska; Ms James MacGregor, Whitefish River Indian Reserve; Dr Peter McNair, University of British Columbia, Victoria; Mr Larry Mao, US Geological Survey; Dr Kaoru Matsuda, University of Arizona, Tucson; Mr Ray Matthews, Great Smoky Mountains National Park; Dr Charles Miksicek, University of Arizona, Tucson; Ms Leuren Moret, Berkeley, California; Dr Randall Morrison, Chaco Culture National Historical Park; Dr John Morton, University of Waterloo, Ontario; Mr James Munson, Effigy Mounds National Monument; Dr Gary Nabhan, Meals for Millions, Tucson, Arizona; New Mexico, Museum of, Laboratory of Anthropology; Dr Virginia Norris, University of California, Davis; John and Carol Overland, Lanesboro, Minnesota; Mr Leo Pahl; Mr John Pehrson, University of California, Riverside; Dr Don Petersen, U.S. Geological Survey, Vancouver, Washington; Dr Robert Perrill, Arizona Sonoran Desert Museum; Dr Thomas Reimchen, University of Alberta, Edmonton; Dr James Risser, Des Moines Register and Tribune; Jim Roach, Western North Carolina Nature Centre; Dr Alan Roelfs, University of Minnesota, St Paul; Faye Russell, Office of Tourism, Baton Rouge; Ms Karen Sausman, Living Desert Reserve; Merton and Elinmore Schlick, Preston, Minnesota; Dr Arnold Schultz, University of California, Berkeley; George Schwab, New Orleans Steamboat Company; Dr A. J. Sharp, University of Tennessee, Knoxville; Mr James Snowdon, Acadia University, Wolfville; Dr John Speth, University of Michigan, Ann Arbor; Dr Patricia Spoerl, US Forest Service, New Mexico; Mr Frank Stratton, Tour West Inc.; Dr Roy Taylor, University of British Columbia; James Temple, Sevierville, Tennessee; Mrs Nancy Turner; Mr Vincent Tutiakof, Ounalaska Corporation; Dr Frank Vasek, University of California, Riverside; Barbara Veloz, Smithsonian Institution, Washington; John and Shirley Welles, Ponchatoula, Louisiana; Mr Richard Whetsell, Oklahoma; Peter White, Great Smoky Mountains National Park; Dr Carroll E. Wood, Arnold Arboretum, Cambridge, Massachusetts; Dr H. Wright, University of Minnesota, Minneapolis; Mrs Sylvia Yeoman, Westmoreland Historical Society, Dorchester, New Brunswick; Mr and Mrs Zadora-Gerloff, Queen Charlotte Islands.

Glyn Davies for the book design, Line and Line for all maps and illustrations, Jennifer Fry for picture research.

Picture Credits

Index

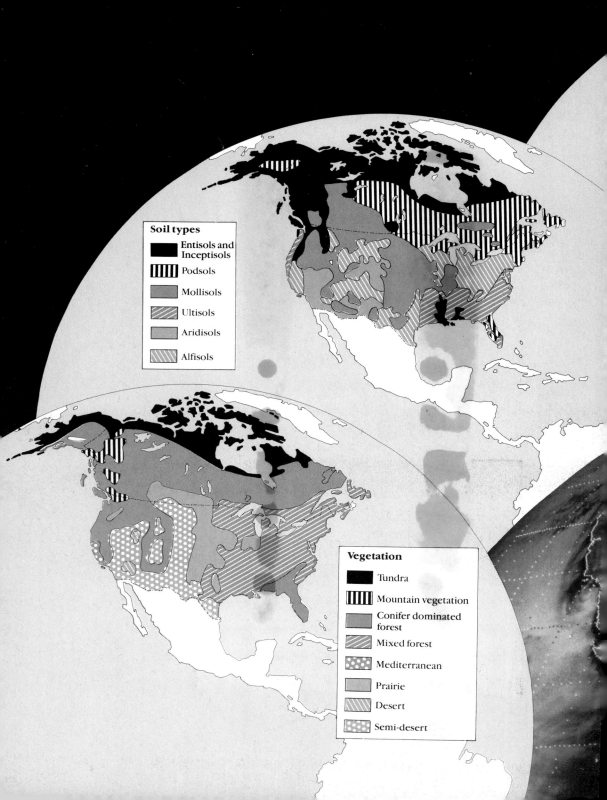

Soil types

- Entisols and Inceptisols
- Podsols
- Mollisols
- Ultisols
- Aridisols
- Alfisols

Vegetation

- Tundra
- Mountain vegetation
- Conifer dominated forest
- Mixed forest
- Mediterranean
- Prairie
- Desert
- Semi-desert